The principal objective of the *Royal College of Radiologists' Clinical Oncology symposia* is to promote multi-disciplinary collaboration for the benefit of cancer patients. Awareness and understanding of advances in a broad range of subjects are essential to ensure the timely and effective application of new techniques in the prevention and treatment of all forms of cancer. Communication of this kind may also stimulate the creation of new ideas that may prove to be of more fundamental relevance to cancer research.

Previously published:

Thyroid Cancer
Edited by William Duncan

Prostate Cancer
Edited by William Duncan

Colorectal Cancer
Edited by William Duncan

Paediatric Oncology
Edited by William Duncan

Lung Cancer
Edited by William Duncan

Ovarian Cancer
Edited by Norman M. Bleehen

Tumours of the Brain
Edited by Norman M. Bleehen

Investigational Techniques in Oncology
Edited by Norman M. Bleehen

Radiobiology in Radiotherapy

Edited by

Norman M. Bleehen

With 61 Figures

Springer-Verlag
London Berlin Heidelberg New York
Paris Tokyo

Professor Norman M. Bleehen, FRCP, FRCR, Honorary FACR,
Cancer Research Campaign Professor of Clinical Oncology;
Honorary Director MRC Clinical Oncology and
Radiotherapeutics Unit, University of Cambridge School of
Clinical Medicine, Department of Clinical Oncology and
Radiotherapeutics, Addenbrooke's Hospital, Hills Road,
Cambridge CB2 2QQ, UK.

ISBN 978-1-4471-1605-9 ISBN 978-1-4471-1603-5 (eBook)
DOI 10.1007/978-1-4471-1603-5

British Library Cataloguing in Publication Data
Radiobiology in radiotherapy.
1. Man. Cancer. Radiotherapy I. Bleehen, Norman M. (Norman Montague),
1930– 616.99'40642
ISBN-13:978-1-4471-1605-9

Library of Congress Cataloging-in-Publication Data
Radiobiology in radiotherapy.
Based on a symposium organized by the Royal College of Radiologists and the International Society for Radiation Oncology; held in London in Feb., 1987.
Includes bibliographies and index.
1. Cancer—Radiotherapy—Congresses. 2. Radiobiology—Congresses. I. Bleehen, Norman M., 1930– . II. Royal College of Radiologists (Great Britain)
III. International Society for Radiation Oncology. [DNLM: 1. Radiation Effects—congresses. 2. Radiation Injuries—congresses. 3. Radiobiology—congresses. 4. Radiotherapy—congresses. WN 610 R1296 1987]
RC271.R3R334 1987 616.99'40642 87–28873
ISBN-13:978-1-4471-1605-9

Softcover reprint of the hardcover 1st edition 1988

Filmset by Wilmaset, Birkenhead, Merseyside

2128/3916 543210

Preface

The ninth annual multidisciplinary symposium on clinical oncology organized by the Royal College of Radiologists was jointly arranged with the International Society for Radiation Oncology. It was held in London in February 1987 and discussed the biological and clinical basis of the effects of radiotherapy. Wherever possible lectures by an experimental scientist were paired with those of a clinical scientist in order to emphasize clinical relevance. It is hoped that this has resulted in a widely balanced view of the subject. The volume presents an updated version of these subjects based on those talks.

After surgery, radiotherapy is the main treatment used in the management of patients with cancer. Its empirical success when first introduced is now backed up by a wealth of laboratory, clinical and experimental experience. New techniques for administering the conventional X-ray therapy have been supplemented by methods which can be used to modify the radiation response. These include changes in dose rate or fractionation, and combined modality treatments including sensitization by drugs or heat. Other types of radiation, such as neutrons and other particles, are also now available which have enhanced physical and biological advantages.

The symposium did not attempt to pass clinical judgement on the various techniques. This would have needed a much larger volume with a critical review of randomized studies, many of which await initiation or completion. What it has attempted to do is to emphasize the importance of the symbiotic relationship that has developed between the physicist, biologist and clinician in this important area of cancer treatment.

I wish to express my thanks to all the speakers for making the symposium a success and agreeing to provide manuscripts for publication. Mr Michael Jackson of Springer-Verlag has continued to give advice and support in the production of this volume. Mrs Jane Farrell of Springer-Verlag has continued to provide her most excellent copy editing. I should also like to thank my personal assistants Ms Leslie Sargeant, Mrs Anne Anderson and Mrs Fiona Clarke for help with the organization of the meeting and the editing of the manuscripts. Finally my thanks are due to Mr A. J. Cowles, General Secretary of the Royal College of Radiologists, and to his staff, who contributed to the organization and management of the symposium.

Royal College of Radiologists
38 Portland Place, London, W1N 3DG

Norman M. Bleehen

May 1987

Contents

1 **Cellular Response to Radiation: Experimental**
N. J. McNally 1

2 **Radiation Response of Tumours: Experimental**
P. R. Twentyman 9

3 **Normal Tissue Response to Radiation: Experimental**
J. Denekamp 17

4 **Measurement of Human Normal Tissue and Tumour Responses**
G. Ross and J. R. Yarnold 31

5 **Late Effects of Radiation in Man**
R. J. Berry 47

6 **Principles of Fractionation in Radiotherapy**
J. F. Fowler 53

7 **Fractionation in the Clinic**
M. V. Williams 59

8 **Radiation Resistance and the Dose-Rate Effect: Experimental**
G. G. Steel and A. Horwich 73

9 **Low Dose Rate in Clinical Radiotherapy**
A. Horwich, P. Blake and G. G. Steel 81

10 **Total Body Irradiation: Normal Tissue Effects**
L. S. Constine and P. Rubin 95

11 Total Body Irradiation: Clinical Aspects
A. Barrett 123

12 Particle Therapy: Physics and Biology
D. K. Bewley 129

13 Clinical Aspects of Particle Therapy
H. M. Warenius and R. D. Errington 137

14 Radiation Sensitizers and Related Compounds: New Approaches to Modification of Radiation Response
G. Adams and I. J. Stratford 153

15 Radiation Sensitizers in Clinical Radiotherapy
S. Dische 165

16 Combined Treatment with Radiation and Anti-Cancer Drugs: Experimental and Clinical Results
H. Bartelink, A. C. Begg, L. Dewit and F. A. Stewart 177

17 Biological Aspects of Hyperthermia
S. B. Field 201

18 Clinical Hyperthermia: Methods and Results
N. M. Bleehen and G. C. W. Howard 209

19 Monoclonal Antibodies in the Diagnosis and Treatment of Cancer
L. W. Brady, D. V. Woo, A. M. Markoe, H. Koprowski and C. T. Miyamoto 233

Subject Index 243

Contributors

G. Adams
Medical Research Council Radiobiology Unit, Chilton, Didcot, Oxon OX11 ORD, UK

A. Barrett
Glasgow Institute of Radiotherapeutics and Oncology, Western Infirmary, Glasgow, UK

H. Bartelink
Netherlands Cancer Institute, Antoni van Leeuwenhoekhuis, Plesmanlaan 121, 1066 CX, Amsterdam, The Netherlands

A. C. Begg
Netherlands Cancer Institute, Antoni van Leeuwenhoekhuis, Plesmanlaan 121, 1066 CX, Amsterdam, The Netherlands

R. J. Berry
Department of Oncology, The Middlesex Hospital Medical School, London W1P 7PN, UK

D. K. Bewley
Medical Research Council Cyclotron Unit, Hammersmith Hospital, Du Cane Road, London W12 0HS, UK

P. Blake
The Royal Marsden Hospital, Fulham Road, London SW3

N. M. Bleehen
University Department and Medical Research Unit of Clinical Oncology and Radiotherapeutics, The Medical School, Cambridge CB2 2QQ, UK

L. W. Brady
Department of Radiation Oncology and Nuclear Medicine, Hahnemann University, Philadelphia PA 19102, USA

L. S. Constine
University of Rochester Cancer Center, Division of Radiation Oncology, 601 Elmwood Avenue, Rochester, New York 14642, USA

J. Denekamp
Cancer Research Campaign, Gray Laboratory, Mount Vernon Hospital, Northwood, Middlesex HA6 2RN, UK

L. Dewit
Netherlands Cancer Institute, Antoni van Leeuwenhoekhuis, Plesmanlaan 121, 1066 CX, Amsterdam, The Netherlands

S. Dische
Marie Curie Research Wing, Regional Radiotherapy and Oncology Centre, Mount Vernon Hospital, Northwood, Middlesex HA6 2RN, UK

R. D. Errington
University of Liverpool Cancer Research Campaign, Department of Radiation Oncology, Clatterbridge Hospital, Clatterbridge Road, Bebington, Wirral, Merseyside L63 4LA, UK

S. B. Field
Medical Research Council Cyclotron Unit, Hammersmith Hospital, Du Cane Road, London W12 0HS, UK

J. F. Fowler
Cancer Research Campaign, Gray Laboratory, Mount Vernon Hospital, Northwood, Middlesex HA6 2RN, UK

A. Horwich
Institute of Cancer Research, Clinical Academic Unit, The Royal Marsden Hospital, Downs Road, Sutton, Surrey SM2 5PT, UK

G. C. W. Howard
University Department and Medical Research Council Unit of Clinical Oncology and Radiotherapeutics, The Medical School, Cambridge CB2 2QQ, UK

H. Koprowski
Wistar Institute, Philadelphia, PA 19104, USA

A. M. Markoe
Department of Radiation Oncology and Nuclear Medicine, Hahnemann University, Philadelphia, PA 19102, USA

N. J. McNally
Cancer Research Campaign, Gray Laboratory, Mount Vernon Hospital, Northwood, Middlesex HA6 2RN, UK

C. T. Miyamoto
Department of Radiation Oncology and Nuclear Medicine, Hahnemann University, Philadelphia, PA 19102, USA

G. Ross
Department of Radiotherapy, The Royal Marsden Hospital, Downs Road, Sutton, Surrey SM2 5PT, UK

P. Rubin
University of Rochester Cancer Center, Division of Radiation Oncology, 601 Elmwood Avenue, Rochester, New York 14642, USA

G. G. Steel
Radiotherapy Research Unit, Institute of Cancer Research, Clifton Avenue, Sutton, Surrey SM2 5PX, UK

F. A. Stewart
Netherlands Cancer Institute, Antoni van Leeuwenhoekhuis, Plesmanlaan 121, 1066 CX, Amsterdam, The Netherlands

I. J. Stratford
Medical Research Council Radiobiology Unit, Chilton, Didcot, Oxon OX11 0RD, UK

P. R. Twentyman
Medical Research Council Clinical Oncology and Radiotherapeutics Unit, Hills Road, Cambridge CB2 2QH, UK

H. M. Warenius
University of Liverpool Cancer Research Campaign, Department of Radiation Oncology, Clatterbridge Hospital, Clatterbridge Road, Bebington, Wirral, Merseyside L63 4LA, UK

M. V. Williams
Radiotherapeutic Centre, Addenbrooke's Hospital, Hills Road, Cambridge CB2 2QQ, UK

D. V. Woo
Department of Radiation Oncology and Nuclear Medicine, Hahnemann University, Philadelphia, PA 19102, USA

J. R. Yarnold
Department of Radiotherapy, The Royal Marsden Hospital, Downs Road, Sutton, Surrey SM2 5PT, UK

1 Cellular Response to Radiation: Experimental

N. J. McNally

Introduction

Radiation is extremely effective at sterilizing cells. The amount of heat energy in a cup of tea would be fatal to a man if it was delivered instead as X-irradiation. This observation, in itself, makes the study of the biological effects of ionizing radiation a fascinating subject. In addition, because radiation is so effective at killing cells and therefore is widely used in the treatment of cancer, there is every reason to devote considerable effort to understanding the mechanism whereby cells are killed and how they can repair and modify radiation damage.

This chapter will be concerned with the lethal effects of radiation on mammalian cells in tissue culture. Rather than reviewing the whole field of cellular radiation biology it will concentrate on three areas, briefly considering some recent developments and aspects of their relevance to radiotherapy. First, it considers the shape of mammalian cell survival curves and some aspects of recovery from radiation damage. Secondly, it reviews the mechanism of the oxygen effect, how oxygen sensitization might be modified and effects at small doses. Finally, the role of tissue culture techniques in the prediction of in vivo tumour radiosensitivity is discussed. The general aim will be to demonstrate that studies of basic mechanisms determining the lethal effects of radiation can, and do, have implications for clinical radiotherapy and will continue to be central to the field of radiation biology.

The techniques of tissue culture and their application to radiation biology have been well described in a number of text books (Elkind and Whitmore 1967; Hall 1978; Freshney 1983) and will not be discussed here.

Survival Curves and Recovery from Radiation Damage

The first mammalian cell survival curve was determined by Puck and Marcus in 1956, using cells from a patient with cervical carcinoma. Since then hundreds, if not thousands, of such curves have been published and all have one of three shapes: exponential with or without a shoulder, or continuously bending downwards with increasing dose. Fig. 1.1 illustrates these three conditions for freshly cultured human diploid fibroblasts (Cox et al. 1977), for V79 Chinese hamster cells (McNally and de Ronde 1976), and for cultured human T1 kidney cells (Barendsen 1962). Various biophysical models have been developed to try and explain the shapes of mammalian cell survival curves. A discussion of these is outside the scope of this chapter, but the interested reader can refer to recent reviews (Blakely and Eddington 1985).

The process whereby the absorption of energy is converted to a biological effect, that is death of a cell, is very complex and involves both physicochemical and biochemical mechanisms. At each stage there are opportunities for the cell to modify the damage through radical and biochemical repair processes. The existence of a shoulder to the survival curve implies some form of damage accumulation process. According to target theory (e.g. Lea 1965) a cell must accumulate a certain number (N) of hits or damaging events before it is killed; any number less than this will be sublethal and the cell will survive. As the radiation dose increases virtually all surviving cells will have received $N-1$ hits and further cell killing will follow "one-hit" kinetics with the survival curve becoming exponential. Target theory says nothing about whether or not cells receiving $N-1$ or less hits can recover from these hits and so behave as though

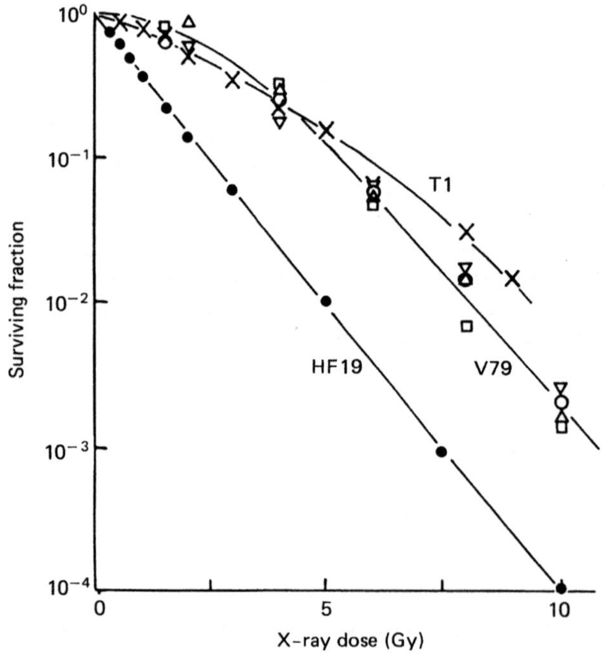

Fig. 1.1. Typical mammalian cell survival curves. *Filled circles*, human fibroblasts (Cox et al. 1977); *open symbols*, V79 Chinese hamster cells (McNally and de Ronde 1976); *crosses*, human T1 kidney cells (Barendsen 1962).

they had not been irradiated at all. However, the demonstration by Elkind and Sutton (1960) that mammalian cells can recover from this sublethal damage (SLD) is explained qualitatively by target theory by postulating that given time the cell can indeed repair the damage caused by these sublethal hits. This ability of the cell to recover from SLD so that, given time (3–6 hours), the survival curve shoulder is restituted, is of profound significance for fractionated radiotherapy and is the principal reason why the total dose has to be increased when it is given as a fractionated course.

The so-called repair theory is an explanation for the shape to the survival curve. This has the repair of radiation damage leading to the restitution of the shoulder of the survival curve as an inherent part of its conceptual framework. It was first proposed by Powers (1962) and subsequently expanded by a number of people (e.g. Alper 1979; Goodhead 1985). According to this theory, the fundamental mode of cell killing is via single-hit kinetics. This induces lesions, perhaps double-stranded breaks in DNA, which can either be fixed in a lethal form within a given time, or are subject to enzymatic repair. The shoulder to the survival curve arises because the repair process becomes saturated as more and more lesions are induced with increasing dose. The initial slope then reflects the induction of irreparable lesions and the final slope the induction of lesions in the absence of repair. The capacity of the repair process is then reflected in the size of the shoulder to the curve. An attraction of such a model is that it provides a framework within which to search for repair mechanisms which could influence shoulder size and perhaps be manipulated to therapeutic advantage. However, as yet the nature of these mechanisms is still poorly understood.

An important question in relation to fractionated radiotherapy is whether or not cells can undergo repeated cycles of damage and repair without damage to their repair mechanism. Virtually all models concerned with the effects of fractionated radiotherapy on normal tissues assume that the repair mechanism is unaffected (e.g. Thames et al. 1982; Fowler 1983). Several in vitro studies have addressed this question and most conclude that the repair capacity is decreased with increasing number of fractions. In their pioneering studies of recovery from sublethal damage, Elkind and Sutton (1960) exposed log-phase V79 Chinese hamster cells to six fractions of 5.05 Gy 22 hours apart. The resulting survival curve was as predicted on the assumption that the repair mechanism was unaffected. A potential complication in the interpretation of these results is that there would have been redistribution of cells through the cell cycle during the fractionated irradiation. In order to overcome this problem, McNally and de Ronde (1976) irradiated V79 cells in the plateau phase of growth when they were in an extended G_1 phase. They found that cells exposed to five or more doses of 2 Gy 6 hours apart had a decreased ability to accumulate SLD (a reduced shoulder to their survival curve) relative to unirradiated cells. In other words the repair/recovery capacity decreased with increasing number of fractions. Zemen and Bedford (1985) and Ngo et al. (1986) came to similar conclusions irradiating C3H 10T$\frac{1}{2}$ cells.

If these observations were also to be true for cells irradiated in vivo, this could have important implications for models used to predict the effect of changing fractionation schedules. All such models assume equal effect per fraction, that is that there is no change in repair capacity during fractionated irradiation (e.g. Douglas and Fowler 1976). Two in vivo studies have specifically addressed this problem. Joiner and Denekamp (1986) compared the effects of 8 or 20 fractions each of 4.5 Gy or less irradiating mouse skin. The effectiveness of each fraction remained constant with no loss of repair capacity in going from 8 to 20 fractions. Fisher and Hendry (1986) irradiated hepatocytes in vivo. They found equal effect per fraction from each of 3 fractions of 8 Gy up to 70 fractions each of 1 Gy assaying colony forming efficiency. It is clearly important to extend these studies to other normal tissues, especially late-responding ones. These are the ones which should be spared as smaller and smaller doses per fraction are used because of the shape of the shoulder to their underlying cell survival curve (Fowler 1983).

The reason for the different response of cells in vitro and skin cells in vivo is not known. It may reflect the generally greater capacity of cells in organized tissues in vivo relative to cells in tissue culture to absorb SLD (McNally et al. 1979). Whatever the explanation, as yet there is no reason to doubt the predictions of models for the response of normal tissues to fractionation based on an equal effect per fraction.

The Oxygen Effect

This section will consider some recent studies on the mechanism of the oxygen effect and on the magnitude of oxygen sensitization at clinically relevant doses. More detailed accounts of the oxygen effect and its potential role in radiotherapy can be found in numerous reviews (e.g. Dische 1978; Alper 1979).

Radiation deposits its energy in biological material through the production of highly reactive free radicals which can react with molecular oxygen to produce organic peroxides capable of "fixing" the damage in a lethal form. In the absence of oxygen these radicals can revert to the undamaged form. This oxygen fixation hypothesis, first proposed by Alper (1956) and Howard-Flanders (1960), provides a framework within which to try and identify the exact chemical species and processes involved in the oxygen effect. In its simplest form the hypothesis can be illustrated as follows:

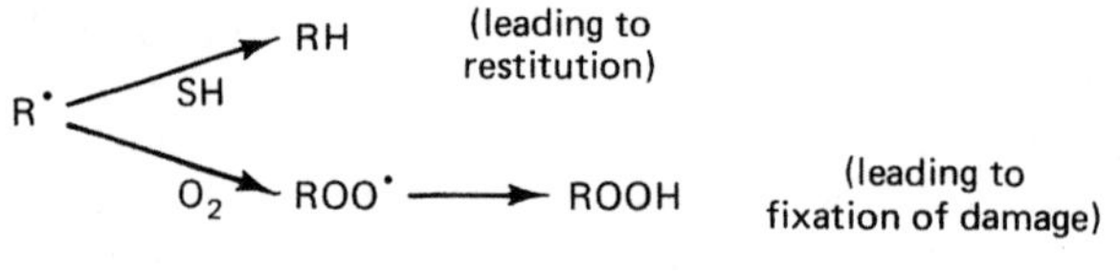

Here $R^{\bullet}$ is the initial target free radical. Electron-affinic sensitizers can mimic oxygen in the fixation of the damage.

The experimental evidence that supports this basic mechanism comes in part from "fast-mixing" experiments whereby hypoxic cells are mixed with various concentrations of oxygen prior to or after irradiation in times of the order of milliseconds. For example, Watts et al. (1978) used a gas explosion technique to show that the full oxygen effect was obtained when oxygen was in contact with cells for only a few milliseconds before irradiation. Recently, Michael et al. (1986) have extended these studies and shown that fast chemical repair of oxygen-dependent free radical damage is completed within 10–20 ms of irradiation. They also showed that, when cells are depleted of their intracellular thiols, the lifetime of the oxygen-dependent damage was extended. However, only about half of the chemical restitution of the damage could be attributable to intracellular thiols. The nature of the remaining repair process is not clear. It is clear that there must be a hierarchy of potentially inactivating reactions which may or may not require oxygen for their fixation. Those not requiring oxygen may be difficult to repair and lead to cell death in hypoxia.

The fact that intracellular thiols are so important in the repair of free radical damage has led to a number of attempts to manipulate the thiol level of cells and so to modify their sensitivity and the extent of sensitization by oxygen and radiosensitizers. The glutathione synthesis blocking agent, buthionine sulphoximine (BSO), has been used extensively in this respect (Griffith and Meister 1979). The basic radiobiological observation can be illustrated by data from Hodgkiss and Middleton (1983). They found that treatment of V79 cells with 100 μM BSO so as to reduce their glutathione content to 10% or less of the control values had no effect on oxic cells but reduced the oxygen enhancement ratio (OER) from 3.2 to 2.7. They further showed that a concentration of the electron-affinic radiosensitizer, misonidazole, which had little effect on normal hypoxic cells, gave a sensitizer enhancement ratio (SER) of about 2 in BSO-treated cells. A similar result has been obtained in hypoxic tumour cells in vivo (Yu and Brown 1984). This obviously raises interesting clinical possibilities regarding the use of radiosensitizers with radiotherapy, and trials are already planned for the use of BSO in man. However, it must be borne in mind that radical repair is an important general mechanism of detoxification in man, and treatments which reduce glutathione in vivo may compromise the ability of cells to combat radical attack from other agents (e.g. alcohol damage).

Experimentally, measurements of cell survival, of DNA damage, and of tumour control probability have all employed radiation doses typically greater than about 4 Gy. In radiotherapy, however, the dose per fraction is usually in the range 1–3 Gy. The question arises, therefore, of whether the OER at low (clinical) doses, or the SER for electron-affinic radiosensitizers, is less than that at higher doses. If the initial slope to the cell survival curve reflects the induction of irreparable damage, one might expect that agents which modify the radiation response would be less effective at low doses. A number of studies have shown that both the OER and the SER (for misonidazole) are less at 2 Gy than at higher doses (McNally and de Ronde 1976; Palcic and Skarsgard 1984; Watts et al. 1986). Typically, the OER is reduced from 3 to 2.0–2.4. This is still considerably greater than unity, so that methods to overcome resistance of hypoxic cells will still be of importance for those tumours in which hypoxic cells are limiting the probability of tumour sterilization.

The protective effect of hypoxia was one of the first observations made in radiation biology. Studies of the oxygen effect continue to provide information on basic mechanisms of radiation action. They have also led to new developments in radiosensitization through the manipulation of cellular thiol levels. These, in turn, are suggesting new approaches to overcome the resistance of hypoxic cells in tumours. Thus, yet again, studies of basic mechanisms lead to potential clinical applications.

Predictive Assays

The previous two sections have been concerned with basic aspects of the cellular response to radiation which, nevertheless, have led to new approaches in radiotherapy. This section is concerned with a very practical application of the techniques of tissue culture and the concepts of cellular radiation biology. It is known that current clinical and pathological methods of staging and classifying cancer are inadequate to predict with reasonable accuracy the response of individual tumours to treatment. Yet the objective of treatment is to kill tumour cells. Can any assessment be made of the radiosensitivity of individual tumour cells which will be of predictive value and hence aid in treatment design? At present it is fair to say that it cannot. However, various tests are being developed, and progress is being made in the measurement of cellular radiosensitivity of human tumours which may allow such predictive assessments on individual patients to be made in the near future.

Since all the clonogenic cells in a tumour need to be killed if the patient is to be cured, there can be no doubt that the intrinsic radiobiological properties of the tumour cells must influence the total dose needed. These properties will include parameters such as the initial slope to the survival curve, the capacity to absorb and repair sublethal or potentially lethal damage, etc. It is doubtful whether the terminal slope (D_0) to the survival curve will be of clinical significance. First, the range of D_0 values that have been measured in human and experimental tumour cell lines is not very great. Secondly, the doses used in clinical practice are too small to be on the exponential part of the cell survival curve.

Two approaches have recently been applied to characterizing tumour cell radiosensitivity in a way that might match clinical tumour radiosensitivity. Weichselbaum (1984) has classified early passages of human tumour cell lines according to their ability to repair radiation damage. The correlation between the radioresistance of the tumour type of origin and of the cultured cells' ability to repair potentially lethal damage was good in a general sense in that cells derived from melanomas and osteosarcomas showed a large capacity to repair radiation damage whereas cells derived from a neuroblastoma did not. These studies point to the importance of repair capability as being a determining factor in radiocurability, but it is doubtful whether such techniques, which require some time (weeks or more) to establish the cell lines, will be of value as predictive assays. They are, rather, identifying mechanisms which are important, the manipulation of which may lead to better strategies for treatment.

An alternative approach first suggested by Fertil and Malaise (1981) and extended by Deacon et al. (1984) has been to analyse the published cell survival curves for established human tumour cell lines. They reasoned that since radiotherapy is usually given in 2 Gy fractions, survival at 2 Gy might be an important indicator of radiosensitivity. In both these studies the authors found that for any given tumour type there was quite a wide range of surviving fractions at 2 Gy, perhaps reflecting the known heterogeneity in response of that tumour type in clinical practice. In addition, however, the mean surviving fraction at 2 Gy correlated well with the known probability of tumour control. For instance, Deacon et al. (1984) found that for neuroblastoma, lymphoma and myeloma cell lines survival at 2 Gy varied from 0.08 to 0.4 with a mean of 0.2.

For melanoma, osteosarcoma and glioblastoma cells the range was from 0.2 to 0.9 with a mean of about 0.54.

If it were possible to make these assessments of cell survival at 2 Gy on short-term cultures from tumours prior to their treatment then such assays might be of real predictive value. It was to this end that Brock et al. (1985) developed a technique applicable to tumour cells assayed within about 14 days of obtaining the biopsy material. They used multi-well plates coated with so-called Cell Adhesive Matrix, a substance which aids cell attachment to the multi-wells. This gives a high success rate for growing cells from a wide range of tumour types. The procedure is to plate the same number (1000–2000) of freshly prepared tumour cells in each of the 24 wells of the plate. They are then exposed to 0–5 Gy in groups of four wells, 24 hours later. Between 12 and 14 days later the cultures are stained with 0.1% crystal violet and the total amount of stain in each dish, which is proportional to cell number, is determined using an optical image analysis system. By relating this to the control value a measure of cell kill is obtained and hence a survival curve for doses up to 5 Gy can be determined. The method is surprisingly sensitive and these authors have been able to confirm the correlations obtained by Fertil and Malaise, and by Deacon and colleagues. This technique has the merit of speed, as an answer can be obtained within 2 weeks of taking the biopsy. More development and characterization of the culture system is needed, but it clearly has the potential to provide information that will aid treatment planning on an individual patient basis.

Conclusions

Three aspects of cellular radiation biology, all of which have the potential to contribute to improvements in the treatment of cancer by radiation, have been reviewed. Without a continuing effort to understand the basic mechanisms whereby radiation kills cells, cells can repair radiation damage and modifiers of radiation response can act, progress in the practical treatment of cancer will be essentially mechanical – improving beam characteristics, development of dynamic conformational radiotherapy etc. Both routes to the improvement of radiotherapy are required.

References

Alper T (1956) The modification of damage caused by primary ionization of biological targets. Radiat Res 5:119–134

Alper T (1979) Cellular radiobiology. Cambridge University Press, Cambridge

Barendsen GW (1962) Dose survival curves of human cells in tissue culture irradiated with alpha-, beta-, 20-kV X- and 200-kV X-radiation. Nature 193:1153–1155

Blakely EA, Eddington MS (eds) (1985) Proceedings of the symposium on heavy charged particles in research and medicine. Radiat Res 104(2) [suppl 8]

Brock WA, Maor MH, Peters LJ (1985) Predictors of tumour response to therapy. Radiat Res 104(2):290–296

Cox R, Thacker J, Goodhead DT (1977) Inactivation and mutation of cultured mammalian cells by aluminium characteristic ultra-soft X-rays. Int J Radiat Biol 31:561–576

Deacon J, Peckham MJ, Steel GG (1984) The radioresponsiveness of human tumours and the initial slope of the cell survival curve. Radiother Oncol 2:317–323

Dische S (1978) Hyperbaric oxygen: the MRC trials and their clinical significance. Br J Radiol 51:888–894

Douglas BG, Fowler JF (1976) The effect of multiple small doses of X-rays on skin reactions in the mouse and a basic interpretation. Radiat Res 66:401–426

Elkind MM, Sutton H (1960) Radiation response of mammalian cells grown in culture. Radiat Res 13:556–593

Elkind MM, Whitmore GF (1967) The radiobiology of cultured mammalian cells. Gordon and Breach, New York

Fertil B, Malaise EP (1981) Inherent cellular radiosensitivity as a basic concept for human tumour radiotherapy. Int J Radiat Oncol Biol Phys 7:621–629

Fisher DR, Hendry JH (1986) Responses of clonogenic hepatocytes to fractionated irradiation. Br J Cancer 53 [Suppl VII]: 298–299

Fowler JF (1983) La Ronde – radiation sciences and medical radiology. Radiother Oncol 1:1–22

Freshney RI (1983) Culture of animal cells: a manual of basic technique. AR Liss, New York

Goodhead T (1985) Saturable repair models of radiation action in mammalian cells. Radiat Res 104(2):58–67

Griffith OW, Meister A (1979) Potent and specific inhibition of glutathione synthesis by buthionine sulfoximine. J Biol Chem 254:7558–7560

Hall EJ (1978) Radiobiology for the radiologist, 2nd edn. Harper and Row, New York

Hodgkiss RJ, Middleton RW (1983) Enhancement of misonidazole radiosensitization by an inhibitor of glutathione biosynthesis. Int J Radiat Biol 43:179–183

Howard-Flanders P (1960) Effect of oxygen on the radiosensitivity of bacteriophage in the presence of sulphydryl compounds. Nature 186:485–487

Joiner MC, Denekamp J (1986) Evidence for a constant repair capacity over 20 fractions. Int J Radiat Biol 49:143–150

Lea DE (1965) Action of radiations on living cells. Cambridge University Press, Cambridge

McNally NJ, de Ronde J (1976) The effect of repeated small doses of radiation on recovery from sublethal damage by Chinese hamster cells irradiated in the plateau phase of growth. Int J Radiat Biol 29:221–234

McNally NJ, George KC, de Ronde J (1979) Recovery from sublethal damage by acutely hypoxic tumour cells in vivo and in vitro. Br J Radiol 52:642–649

Michael BD, Davies S, Held KD (1986) Ultrafast chemical repair of DNA single and double-strand break precursors in irradiated V79 cells. In: Simic MG, Grossman L, Upton AC (eds) Mechanisms of DNA damage and repair. Plenum Press, New York and London, pp 89–100

Ngo QH, Youngman Y, Shozo S, Iptaia K, Iliakes G (1986) Evidence for reduced capacity for damage accumulation and repair in plateau phase C3H 10T½ cells following multiple doses of γ-rays. Radiat Res 106:380–395

Palcic B, Skarsgard LD (1984) Reduced oxygen enhancement ratio at low doses of ionizing radiation. Radiat Res 100:328–339

Powers EL (1962) Considerations of survival curves and target theory. Phys Med Biol 7:3–28

Puck TT, Marcus PI (1956) Action of X-rays on mammalian cells. J Exp Med 103:653–666

Thames HD, Withers HR, Peters LJ, Fletcher GH (1982) Changes in early and late radiation response with altered dose fractionation: implications for dose survival relationships. Int J Radiat Oncol Biol Phys 8:219–226

Watts ME, Maughan RL, Michael BD (1978) Fast kinetics of the oxygen effect in irradiated mammalian cells. Int J Radiat Biol 33:195–199

Watts ME, Hodgkiss RJ, Jones NR, Fowler JF (1986) Radiosensitization of Chinese hamster cells by oxygen and misonidazole at low doses. Int J Radiat Biol 50:1009–1021

Weichselbaum RR (1984) The role of DNA repair processes in the response of human tumours to fractionated radiotherapy. Int J Radiat Oncol Biol Phys 10:1127–1134

Yu NY, Brown JM (1984) Depletion of glutathione in vivo as a method of improving the therapeutic ratio of misonidazole and SR–2508. Int J Radiat Oncol Biol Phys 10:1265–1269

Zeman EM, Bedford JS (1985) Loss of repair capacity in density-inhibited cultures of C3H 10T½ cells during multifraction irradiation. Radiat Res 104:71–77

2 Radiation Response of Tumours: Experimental

P. R. Twentyman

Introduction

In this chapter our knowledge regarding cellular radiation response and the factors which modify it will be related to the volume changes and probability of control of irradiated solid tumours. After a discussion of the different cell populations present within solid tumours the cell population kinetics of the neoplastic cells will be considered in more detail. The influence of factors related to the three-dimensional geometry of the tumour, particularly hypoxia, will be considered, and also the role of the tumour vasculature in radiation response. Repair of sublethal damage (SLD) and potentially lethal damage (PLD) will be dealt with and finally the relationship between the various end-points of tumour radioresponsiveness will be discussed.

Cell Populations Within Tumours

Within the volume of a solid tumour there are many elements in addition to the neoplastic cell population. The tumour depends for its growth upon a developing blood supply and hence, at any given time, will contain capillary endothelial cells and circulating blood cells. In addition fibroblasts and macrophages will usually be present in significant numbers. In some tumour types, various elements of extracellular matrix will occupy a considerable proportion of the tumour volume. The neoplastic cell population is, itself, frequently heterogeneous with regard to a wide range of parameters. Although the evidence suggests that most human tumours are monoclonal in origin (i.e. they represent the progeny of a single neoplastic cell), it is accepted that clonal heterogeneity results from a genetic instability present during tumour progression (Nowell 1976). From some human tumour cell lines it has been possible to isolate distinct subpopulations of cells with genetically stable differences in intrinsic radiosensitivity (Leith et al. 1982). If this is true for human tumours in situ it would predict that regrowth of a tumour following an initial, subcurative course of therapy would occur from the most resistant cells originally present and hence the recurrent tumour would itself be more radioresistant. This is in accord with clinical experience.

Within many human tumours considerable differentiation is seen and it thus appears likely that many of the neoplastic cells will have differentiated beyond the point where the capacity for extensive proliferation is lost. If this process is irreversible, these cells will be irrelevant with regard to the ultimate radiocurability of the tumour. From this point of view the therapist will be concerned only with tumour stem cells. Stem cells are those cells which retain the capacity for unlimited self-renewal and hence the ability to regrow the tumour following a subcurative course of radiation. The existence of such stem cells within tumours has been inferred by analogy with normal renewal tissues such as the bone marrow or intestinal epithelium. Direct evidence that they exist as an identifiable subpopulation distinct from the majority of the neoplastic cells is, however, lacking. It is possible to make a cell suspension and grow colonies in culture from single cells of both experimental animal tumours and human tumours. The in vitro environment is, however, very different from the environment within a solid tumour and so it would be wrong to equate the proportion of in vitro clonogenic cells with the stem cell content of the in vivo tumour. It does, however, seem likely from the evidence of in vitro cloning data and from data regarding radiocurability of clinical tumours, that the proportion of stem cells in many human tumours is very low.

Proliferation Kinetics of Neoplastic Cells

The radiosensitivity of cells varies at different phases of the cell cycle and there are also changes in sensitivity which occur as cells leave the cell cycle and enter a resting phase. It is therefore necessary to consider these factors in the context of solid tumours. Additionally, in order to understand the volume changes which occur following irradiation of a solid tumour, it is necessary to have a knowledge of the factors which contribute to the unperturbed rate of tumour growth.

The viable cells within the malignant cell population have the potential for extremely rapid growth. The cell cycle of some experimental rodent tumours can be as short as 12–18 hours and the available data for human tumours suggests a median intermitotic time of 2–3 days (Steel 1977). However, two additional factors mean that the actual doubling time of tumours is very much slower than these values would indicate. Firstly, many of the cells are not, in fact, actively dividing, that is they do not comprise part of the growth fraction. Secondly, there is a constant cell loss occurring from the population. The growth fraction for experimental tumours is typically in the range 50%–80% and the cell loss factor (i.e. rate of cell loss as a percentage of the rate of cell production) can be as low as 10% or as high as 90%. For human solid tumours, growth fractions of

20%–70% are typically estimated whilst cell loss factors can be as low as 40%, as for sarcomas, or in excess of 90%, as for colorectal carcinomas or undifferentiated bronchial carcinomas (Steel 1977). The net result of these two factors means that volume doubling times of human tumours usually lie in the range of 30–150 days rather than the 2–3 days which is the cell cycle time.

The experimental observations regarding variations in cellular radiosensitivity at different phases of the cell cycle and between dividing and non-dividing cells are difficult to extrapolate to the situation in solid tumours. It remains unclear whether re-assortment of cells around the cycle contributes to the improved effectiveness of fractionated regimes of radiotherapy. There is, however, evidence that human tumours with a high growth fraction tend to be more easily controlled by radiotherapy (Tubiana 1983). If this is due to a relative radioresistance of non-proliferating cells this could either be an intrinsic difference, or be determined by the ability of such cells to recover from PLD (see below). Alternatively it could indicate a tendency of tumours having a low growth fraction to contain a relatively high proportion of hypoxic cells.

Tumour Vasculature

The growth and therapeutic response of tumours are dependent upon the relationship between the neoplastic cells and the vasculature which supplies them with oxygen, nutrients, etc. It was shown many years ago that mouse tumour cells inoculated into an irradiated area of skin grew less well than those inoculated into an unirradiated site (Hewitt and Blake 1968). This was attributed to the compromised ability of the capillaries in the irradiated site to develop into a tumour vasculature. It is currently believed that one mechanism by which hyperthermia treatment can exert a preferential effect against tumours is by selective damage to tumour blood vessels. Radiation damage to blood vessels is thought to influence the pattern of regrowth of irradiated solid tumours. It is frequently observed in experimental tumours that the initial stage of rapid regrowth following irradiation slows considerably as the tumour passes its treatment volume and then requires the development of new capillaries (Fig. 2.1) (Thomlinson and Craddock 1967; Curtis and Tenforde 1980).

Role of Tumour Geometry and Hypoxia

Thomlinson and Gray (1955) first postulated that tumours contain a proportion of hypoxic cells located beyond the diffusion range of oxygen from the developing vasculature. This idea was based on histological observation that, in some solid tumours, cords of viable tumour cells are seen surrounding blood vessels whereas regions of necrosis are seen at distances of greater than 150–200 μm from the nearest vessel. This distance agrees well with the calculated value of oxygen diffusion distance. It therefore seemed likely that cells would be hypoxic for some period of time before losing viability. If such hypoxic cells do

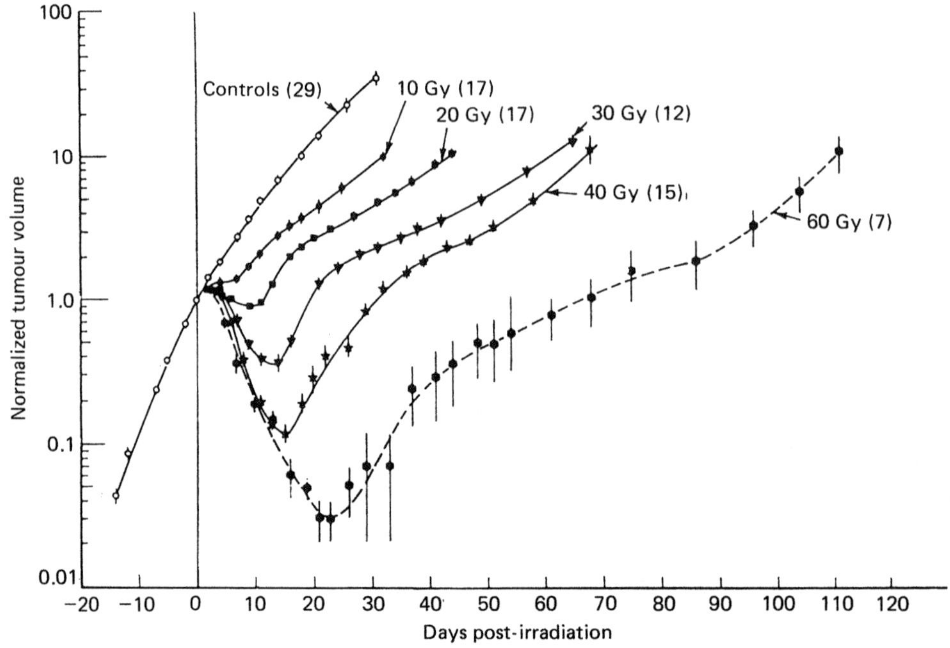

Fig. 2.1. Volumes of R1/LBL rat tumours as a function of time after irradiation with various doses of 220-kV X-rays. Volumes have been normalized to unity on the day of irradiation. Numbers of tumours irradiated at each dose are in parentheses. (After Curtis and Tenforde 1980, with permission.)

exist in solid tumours, they will have profound implications for tumour radiocurability as hypoxic cells are approximately 3 times less sensitive than fully oxygenated cells to X or γ radiation. The radiation response curve for cells from a tumour with such a hypoxic fraction will initially be similar to that for oxic cells. At higher doses, however, most oxic cells will have been killed and the curve will become parallel to that for hypoxic cells. It is the extrapolation of this terminal portion of the curve which will give a prediction of the dose required for tumour control. Even for tumours with very small hypoxic fractions (say 1%) the predicted dose is much higher than that for tumours with a fully oxic population (Hall 1978).

Many human tumours have been shown to contain considerable regions of necrosis and it has been assumed that they therefore contain hypoxic cells. It is not necessarily true, however, that cells lose viability only because of diffusion-limited hypoxia and hence must spend a period of time in a hypoxic, yet viable, state. It could also happen that cells perhaps die due to lack of glucose or build-up of toxic metabolites before becoming sufficiently hypoxic to modify their radiosensitivity. In this case, areas of necrosis may not be significant and regions of hypoxia may consist entirely of dead cells and debris. The balance of evidence strongly suggests, however, that at least some human tumours contain a population of viable, hypoxic cells. An alternative pathway to hypoxia could occur if blood vessels within a tumour undergo cyclic occlusion. This would lead to periods of transient hypoxia in some regions. Such an effect has been demonstrated in certain experimental systems but its clinical significance is unclear. With regard to hypoxia, however, it should be remembered that cells which are hypoxic during one dose of a course of fractionated radiotherapy may not be hypoxic at the time of subsequent doses. Reoxygenation has been shown to occur with widely differing time courses over the week following single dose irradiation of animal tumours (e.g. van Putten and Kallman 1968; Howes 1969), although the mechanism remains unclear. It is, however, likely to be a major factor determining optimal fractionation in clinical radiotherapy.

Repair of Sublethal and Potentially Lethal Damage

The ability of tumour cells to repair SLD or PLD will influence both optimal fractionation and the ultimate radiocurability of the tumour. There is no doubt that both SLD (as demonstrated by split-dose irradiation experiments) and PLD (as demonstrated by delayed subculture following irradiation of cells in the plateau phase of growth) occur in some lines of cultured human tumour cells over a period of 6–8 hours (Weischelbaum and Little 1983). Those authors have found a clear relationship between the ability of cells from different types of human tumour to repair PLD and the radiocurability of that tumour type. Other workers, however, have shown a strong correlation between radioresponsiveness of different human tumours and the initial slope of the in vitro radiation response curve (Fertil and Malaise 1981). A number of studies of human tumour cells growing as solid tumour xenografts in immune-deprived mice have also demonstrated the existence of PLD repair (Chavaudra et al. 1981).

Much attention has been focused on the question of the relative abilities of oxic and hypoxic cells to carry out SLD and PLD repair. The balance of evidence would suggest that hypoxic cells have a reduced ability to repair SLD whilst being particularly efficient at carrying out PLD repair. It must, however, be remembered that normal tissues can also repair both SLD and PLD. Optimal fractionation will therefore be dependent upon a differential in SLD repair between tumour and normal tissue in either its extent or timing. Furthermore, strategies designed to inhibit repair of radiation-induced PLD will only be useful if differential inhibition of repair in the tumour is achieved.

End-points of Radiation Response

Following irradiation of a solid tumour many complex changes occur within the various cell populations which make interpretation of gross volume changes extremely difficult. The main effect of the irradiation will be to damage the reproductive capacity of dividing cells. This applies not only to the stem cell population of cells with unlimited reproductive potential but also to cells which have proceeded so far along the differentiation pathway that only limited reproductive potential remains. The extent of this removal of proliferative potential (i.e. cell kill) will be determined during the time of irradiation and immediately afterwards during PLD repair. The influence upon tumour volume will, however, take days or weeks to become clear. As cell production grinds almost to a halt following irradiation, the rate of tumour shrinkage will be determined primarily by the normal rate of cell loss. This implies that the initial shrinkage rate will not depend to any great extent upon the radiation dose and this has been shown to be true in a number of studies upon experimental tumours (e.g. as in Fig. 2.1). Whilst this shrinkage is occurring, however, any surviving stem cells of the tumour will have begun to repopulate. But if these cells are only a small proportion of the total cells, very extensive repopulation will have to occur before any influence upon tumour volume is seen. In some instances in experimental tumour systems several orders of magnitude of stem cell repopulation can occur whilst the tumour is still shrinking. Eventually, though, the rate of cell production will exceed the natural rate of cell loss and the volume will begin to increase.

Although in some studies of human tumours a correlation has been seen between the early rate of tumour regression and the eventual outcome, this relationship too is extremely complex. It has been calculated that under some conditions the dose required to bring about complete regression of a 1-g tumour may be only one-third of the dose necessary for long-term control (Tubiana 1983).

Conclusion

It should be clear from the discussion here that the complexity of population heterogeneity, cell kinetics, tumour geometry and dynamic processes such as reoxygenation make direct extrapolation from studies with cells in vitro to solid tumours in the clinic extremely difficult. An appreciation of these complexities will, however, assist in an understanding of why much development in clinical radiotherapy has been empirical in nature and why significant observations in the laboratory do not lead necessarily to improvements in treatment of patients.

References

Asterisks indicate major reviews which are recommended for further background reading.

Chavaudra N, Guichard M, Malaise EP (1981) Hypoxic fraction and repair of potentially lethal radiation damage in two human melanomas transplanted into nude mice. Radiat Res 88:56–68

Curtis SB, Tenforde TS (1980) Assessment of tumour response in a rat rhabdomyosarcoma. Br J Cancer 41 [Suppl IV]:266–274

Fertil B, Malaise EP (1981) Inherent cellular radiosensitivity as a basic concept for human tumor radiotherapy. Int J Radiat Oncol Biol Phys: 7:621–629

*Hall EJ (1978) Radiobiology for the radiologist, 2nd edn. Harper and Row, New York

Hewitt HB, Blake ER (1968) The growth of transplanted murine tumours in pre-irradiated sites. Br J Cancer 22:808–824

Howes AE (1969) An estimation of changes in the proportion and absolute numbers of hypoxic cells after irradiation of transplanted C3H mouse mammary tumours. Br J Radiol 44:299–304
Leith JT, Dexter DL, De Wyngaert JK et al. (1982) Differential responses to X-irradiation of subpopulations of two heterogeneous human carcinomas in vitro. Cancer Res 42:2556–2561
Nowell PL (1976) The clonal evolution of tumour cell populations. Science 194:23–28
*Steel GG (1977) Growth kinetics of tumours. Clarendon Press, Oxford
Thomlinson RH, Craddock EA (1967) The gross response of an experimental tumour to single doses of X-rays. Br J Cancer 21:108–123
Thomlinson RH, Gray LH (1955) The histological structure of some human lung cancers and the possible implications for radiotherapy. Br J Cancer 9:539–549
*Tubiana M (1983) The causes of clinical radioresistance. In: Steel GG, Adams GE, Peckham ML (eds) The biological basis of radiotherapy. Elsevier, Amsterdam, pp 13–33
van Putten LM, Kallman RF (1968) Oxygenation status of a transplantable tumor during fractionated radiotherapy. J Natl Cancer Inst 40:441–451
*Weichselbaum RR, Little JB (1983) X-ray sensitivity and repair in human tumour cells. In: Steel GG, Adams GE, Peckham MJ (eds) The biological basis of radiotherapy. Elsevier, Amsterdam, pp 113–121

3 Normal Tissue Response to Radiation: Experimental

J. Denekamp

Introduction

The effectiveness of radiotherapy in treating cancer is limited by the reaction of the normal tissues inevitably included in the beam. Thus the prescribed dose is actually determined by normal tissue tolerance rather than by the type or size of tumour that is being treated. This is recognized in the many fractionation formulae, such as nominal standard dose (NSD) or cumulative radiation effect (CRE) calculations, for determining the total dose in two schedules that will give equivalent normal tissue damage (isoeffective doses). The challenge is then to find methods of changing radiotherapy practice in a way that allows an increase in the effect on the tumour whilst maintaining an isoeffective exposure of normal tissues. This is the aim in chemical modification of response with tumour radiosensitizers or normal tissue radioprotectors, with accelerated or hyperfractionation and with adjuvant therapy involving cytotoxic drugs or hyperthermia.

The foregoing paragraph implies that normal tissues behave in a uniform way, and that tumours behave in a different but also uniform way. Neither of these implications is borne out by the experimental data. Tumours are widely acknowledged to be heterogeneous populations, but it is perhaps less frequently acknowledged that there are fundamental differences between different normal tissues. The common custom of scoring acute skin reactions (both in the clinic and in mouse experiments) and extrapolating from these in order to predict whether a new treatment will have later life-threatening side-effects has recently been shown to be quite misleading.

As regards fractionation effects there seems to be a similarity between many acutely reacting tissues; these effects differ from those in late reacting tissues (Withers et al. 1982). This difference seems to relate to the proliferation rate of the clonogenic cells in each tissue, which in turn relates to the natural rate of removal by wear and tear, i.e. the turnover time (Denekamp 1982, 1986). The vascular architecture is known to be important in the response to drugs, both directly by affecting drug distribution and also more indirectly by influencing the precise distribution of oxygen tensions, and perhaps of pH. Vascular architecture also affects a tissue's response to hyperthermia, as the ability to vary blood flow – by capillary dilation, by opening reserve vessels and by shunts – is of major importance in determining the ability to dissipate applied heat. This is generally more effective in normal tissues than in tumours.

Lethality Studies

Whole body irradiation leads to death if a critical dose is exceeded. The dose giving 50% survival is defined as the LD_{50}. However, no single LD_{50} value can

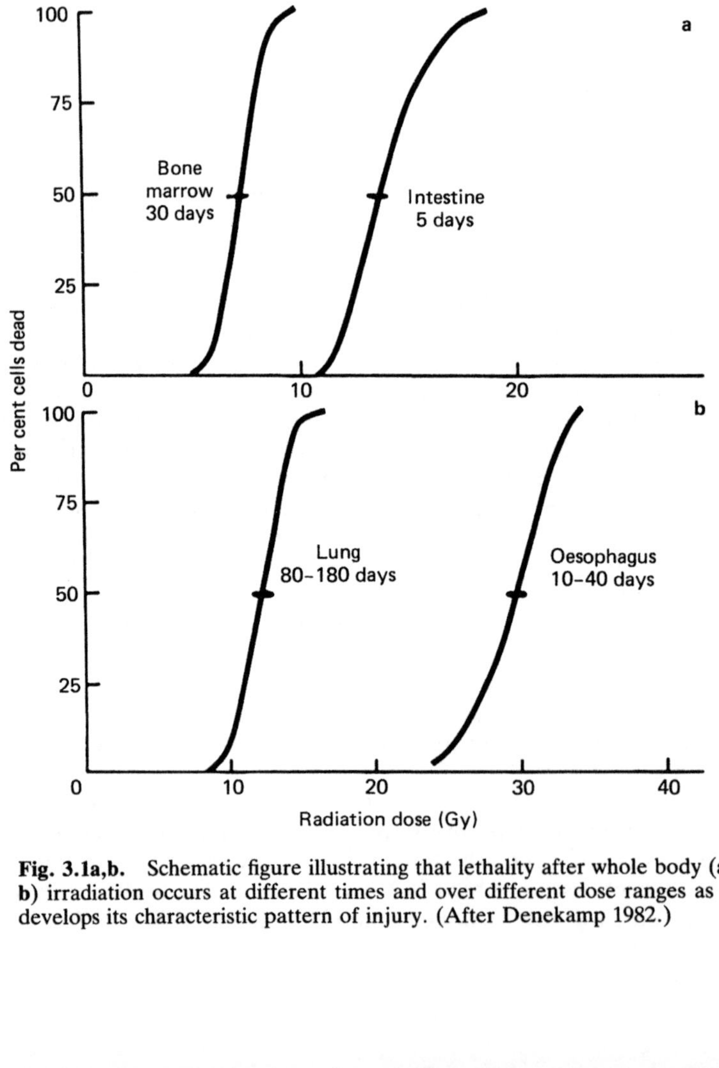

Fig. 3.1a,b. Schematic figure illustrating that lethality after whole body (**a**) or localized (thoracic: **b**) irradiation occurs at different times and over different dose ranges as each tissue in the beam develops its characteristic pattern of injury. (After Denekamp 1982.)

be quoted without defining both the tissue at risk and the period of observation. Approximately 100 Gy are needed to cause instance death in mice (from damage to the central nervous system). A much lower LD_{50} is measured (approximately 13 Gy) if the animals are observed for 1 week; the deaths at between 3 and 5 days result from intestinal injury (Fig. 3.1a). Prolonging the observation period to 1 month allows mice dying of bone marrow dysfunction to be detected, and the LD_{50} falls to 7 or 8 Gy (Fig. 3.1).

The same type of phenomenon is seen after localized irradiation of, for example, the thorax (Fig. 3.1b). Oesophageal damage leads to deaths within 40 days and the LD_{50} is about 30 Gy. At much later times (3–6 months) the surviving mice develop lung damage and the LD_{50} falls to about 12 Gy.

These examples of single dose LD_{50} values emphasize that the tolerance dose differs markedly from one tissue to another, and that the time of expression of injury is characteristic for each tissue. There appears to be a threshold dose below which radiation is safe, with a steep response beyond that, leading from 0 to 100% deaths over a narrow dose range of only 10% or 20%. In conventional clinical practice this threshold, or tolerance dose, is fairly well established for many tissues (Rubin and Casarett 1968). It is well recognized that superficial epithelial damage is expressed rapidly, during or shortly after the course of therapy, whereas deeper tissues such as lung, kidney and spinal cord express their damage months or even years later. Using the immediate reaction to predict the severity of late reactions may be useful for detecting genetically abnormal, radiosensitive individuals (for example those with ataxia telangiectasia). However, in normal individuals it is not safe to predict the altered late response to a change in fractionation schedule or to a combined modality therapy from early reactions (Denekamp 1986). For this reason it is important to understand the biological principles underlying normal tissue responses.

Mitotic Death

Ionizing radiation can damage every molecule in a cell, but DNA is by far the most sensitive target molecule. Base damage and/or strand breaks are inflicted, which can be removed by a variety of repair enzymes. Some of this repair restores the chromosomal material perfectly to its original state, but imperfect repair (misrepair) may leave altered genetic codes or even lead to reassortment between chromosomes by exchange of chromatids or chromosomal fragments (Hall 1978; Alper 1979; Curtis 1986). If unrepaired or misrepaired lesions persist, the cell can usually function perfectly well. However at the time of a subsequent mitosis the DNA is unevenly distributed and an under- or overdosage of genetic material leads to cell death. Thus, after irradiation cell death is not detectable until the cells attempt to divide (with the exception of the lymphocytes and gonads). After low doses cells may succeed in dividing once or twice, and then fail. After higher doses the genetic damage is more severe, and death occurs at the first division.

This mitotic death is the reason why damage is expressed at different times in different tissues. In the intestinal mucosa the mean intermitotic interval, often called the cell cycle time, for crypt cells is about 12 hours. After 10–15 Gy dead cells are detectable within hours in the crypts. Normal wear and tear on the villus removes differentiated cells, which are not replenished as normal from the crypt. Within 3–4 days the villus become flattened and totally denuded. Fluid and electrolyte balances are then disturbed and bacteria gain access to the bloodstream. Thus failure of crypt cell proliferation ultimately leads to death from dehydration and systemic infection within a week. It has been demonstrated, using an in vivo colony forming assay, that the critical cells are in the proliferative compartment, that is the crypts (Hornsey 1973).

Depletion of proliferative cells in other tissues is also occurring. In the bone marrow, where the stem cells have a cell cycle time of about 24 hours, cell death is manifest within a day or two, but the circulating cells derived from the marrow before irradiation continue to function. Thus white cell depletion becomes important over a period of days to weeks, whereas the erythrocyte count remains essentially normal. The fall in white blood cell count leads to a failure to combat systemic infections even if the gastrointestinal epithelium has completely recovered, and deaths occur over 2–4 weeks.

Table 3.1. Correlation of cell turnover time and expression of radiation injury

Tissue	Cell cycle time (hours)	Labelling index (%)	Turnover time (days)	Time of expression of radiation injury (days)	Time of onset of compensatory proliferation (days)
Intestine	12	20	3–4	2–5	2
Skin	100	5–10	15–20	10–30	10–15
Oesophagus	Not known	5–7	4	4–20	3
Lung	Not known	~0.1	30–100	90–180	> 40
Kidney	Not known	~0.1	100–200	100–200	> 40
Bladder	Not known	0.1	350	180–350	>180

A similar correlation of time of expression of injury with turnover time can be demonstrated for other tissues (Table 3.1). In general the tissues with slow turnover show very little uptake of the DNA precursor tritiated thymidine (^{3}HTdR). These tissues show late expression of injury. Tissues that turn over rapidly have a high ^{3}HTdR labelling index and express their damage early. An important consequence of this behaviour is the pattern of compensatory proliferation in response to radiation damage. Tissue depletion is recognized within 2 days in the intestine and a burst of even faster proliferation occurs to heal the tissue, with the cell cycle time shortening to 6 hours. In skin, where the cell cycle time is normally 100 hours, depletion is recognized after 1–2 weeks and the cycle then shortens in the healing phase to less than 1 day. In bladder, kidney, lung, spinal cord and other slowly proliferating tissues the ^{3}HTdR labelling index is very low and no detectable proliferative burst occurs for months (Denekamp and Fowler 1977).

Phase Sensitivities

Cells in different stages of the cell cycle have different radiosensitivities (Sinclair 1972). This is illustrated in Fig. 3.2. Generally cells in mitosis are most sensitive and those in DNA synthesis (the S phase) most resistant, while cells in G_1 and G_2 phases have intermediate sensitivities. In cells with a long G_1 phase there may be several different types of sensitivity at different parts of that phase. When an asynchronous, that is normal, population is irradiated the cells in the sensitive phase are most readily killed, whilst those in the resistant phase predominate among the survivors. The doomed and surviving cells cannot be distinguished at this point, but functional tests such as the response to another dose of radiation show that the population shifts from having a greater radioresistance to an increasing sensitivity to radiation as the survivors move through the cycle towards the mitosis at which the doomed cells will die. Since rapidly and slowly proliferating tissues have different fractions of the population in S phase and mitosis the overall sensitivity can be expected to differ. Moreover, rapidly proliferating cells will quickly reassort into more sensitive phases after a first dose and will soon redistribute into an essentially asynchronous state again. Slowly proliferating cells may progressively become more and more resistant to a series of fractionated doses because their reassortment will be proportionately slower.

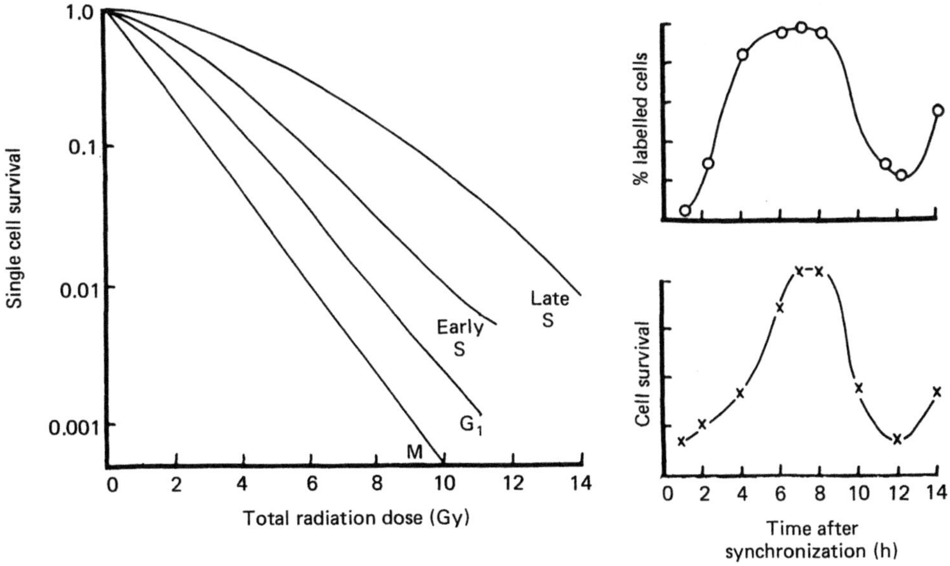

Fig. 3.2. Variation in the sensitivity of V79 cells in vitro when they are irradiated as synchronized cells in different parts of the cell cycle (data from Sinclair 1972). The left-hand graph shows survival curves in four different phases. The right-hand graphs show that after a fixed dose of 7.1 Gy the fraction of cells in the S phase after mitotic shake off coincides with the time of the highest surviving fraction. (After Denekamp 1986.)

Mitotic Delay

One of the non-lethal aspects of radiation damage is that it delays the progression of cells around the cell cycle. The most marked delay is seen in G_2 phase, where cells are arrested before attempting mitosis. Indeed, so many cells accumulate here that the mitotic index, that is the fraction of cells in mitosis, can fall to zero for a time that is dose dependent. These cells then enter mitosis in a cohort and an increased mitotic index may be observed. The time from irradiation to recovery of the mitotic index is termed the mitotic delay, and it is dose dependent (Fig. 3.3). Fig. 3.3a shows plucked and unstimulated skin (with cell cycle times of 47 and 100 hours respectively) and two cell lines irradiated in vitro. Each system shows an increasing mitotic delay with dose which is

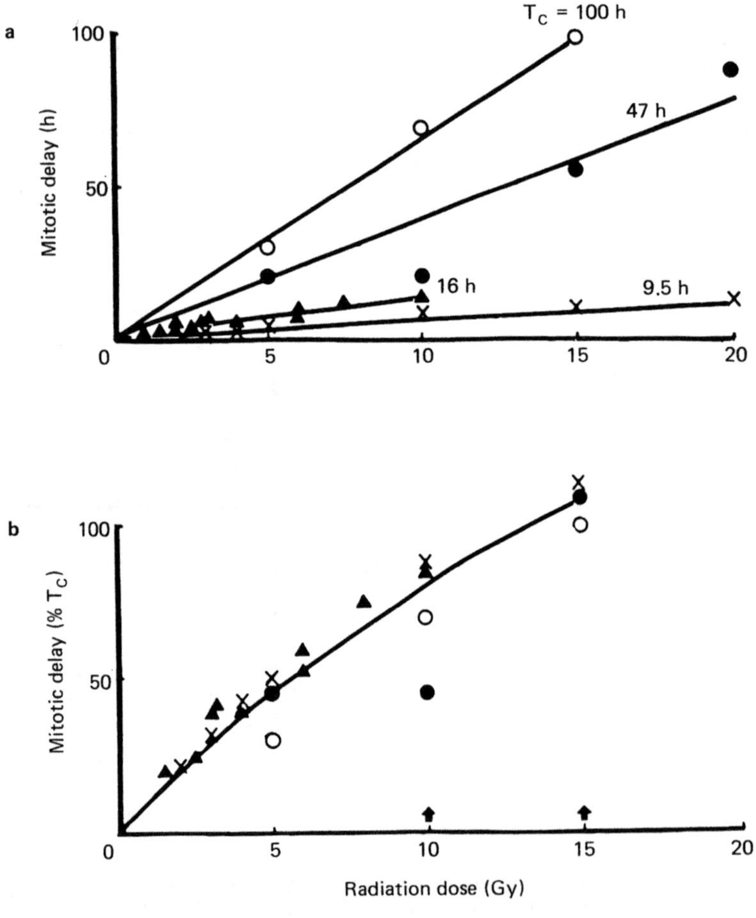

Fig. 3.3. **a** Mitotic delay is proportional to dose in four different systems, but the delay is longer in the cells with the longest cell cycle time (T_C). **b** If the data are expressed as a percentage of the cell cycle time the four sets of data can be fitted by a common line. (After Denekamp 1986.)

approximately linear, but differs from system to system. It ranges from about 10 to 100 hours delay for a dose of 15 Gy. If the data are normalized and expressed as a fraction of the cell cycle time they fit reasonably well to a single line, with a delay equivalent to approximately one cell cycle being induced by 15 Gy (Denekamp 1982).

This arrest of the cells in G_2 phase will increase the synchrony induced by differential phase sensitivities and reassortment. The cell cycle dependence of the delay means that in slowly proliferating tissues doomed cells will take even longer than might have been expected to attempt mitosis, fail, and express their loss of clonogenic potential.

The Cellular Basis of Tissue Response

It is clear that radiobiologists think of normal tissue injury (and of local tumour eradication) in terms of sterilizing the clonogenic precursor cells, which then fail to maintain the supply of functional differentiated cells. There is ample evidence that this is an important component of the response. So many normal tissues can now be assayed by clonogenic techniques, which allow colonies from individual survivors to be recognized, that a monograph has been published listing these assays (Potten and Hendry 1985). But the emphasis on the quantitation of cell survival – in the radiobiological sense of retaining proliferative capacity – should not blind us to the fact that tissues have varying degrees of complexity and of interdependence of different cell types (see papers in Hendry et al. 1986). A simple epithelium, such as skin, can recover and function well even if the clonogenic cells are depleted to 0.1% or even 0.01%. However, in the kidney, where a functioning nephron is a self-contained unit of about 1000 cells, it would clearly be disastrous if less than 1 cell per nephron survived. In intestine, individual crypts contain about 250 cells, of which 150 are proliferating, but if all these are lost adjacent surviving crypts may enlarge or even subdivide to form two new crypts. In lung, several different cell types are needed to maintain the fully functioning alveolus (surfactant-producing cells, epithelial cells, endothelial cells) and the appropriate balance of these must be maintained. Thus although cell killing is the basis of tissue injury the complex biology of each tissue will determine the impact that a certain level of cell depletion has on that particular organ's function. These factors all combine to determine the overall tolerance dose (see papers in Hendry et al. 1986).

Practical Applications of Radiobiology

Over the last decade techniques have been developed that allow much better quantitation of normal tissue injury in a wide range of tissues. Because of the excellent resolution that can now be obtained, radiobiologists have been able to study the influence of different radiations, different chemomodifiers and different fractionation schedules in great detail. Two examples of fractionation are shown in Figs. 3.4 and 3.5. In these examples an identical style of experiment has been used, with the dose progressively being subdivided into twice as many fractions for each curve. The loss of function (i.e. the observed "effect") when plotted against dose gives well-defined dose–response curves (Figs. 3.4a and 3.5a). The sparing effect of fractionation is similar for skin and kidney from 1 to 16 fractions, but beyond that the two tissues differ in their response. Little further sparing is seen in skin when the dose is subdivided into 32 or 64 smaller fractions, but a great deal of extra sparing is seen in kidney. When the data are translated into reciprocal dose plots, back extrapolation gives a measure of the α/β ratio, which reflects the amount of reparable to irreparable injury. If the value for α/β is low, e.g. 3 Gy for kidney, this implies a very curvy response at low doses for the underlying survival curve (Fig. 3.5c). If the α/β value is high, e.g. 10 Gy for skin, it implies an almost linear killing with dose at low dose levels (Fig. 3.4c). (For further details of these analyses consult Douglas and Fowler

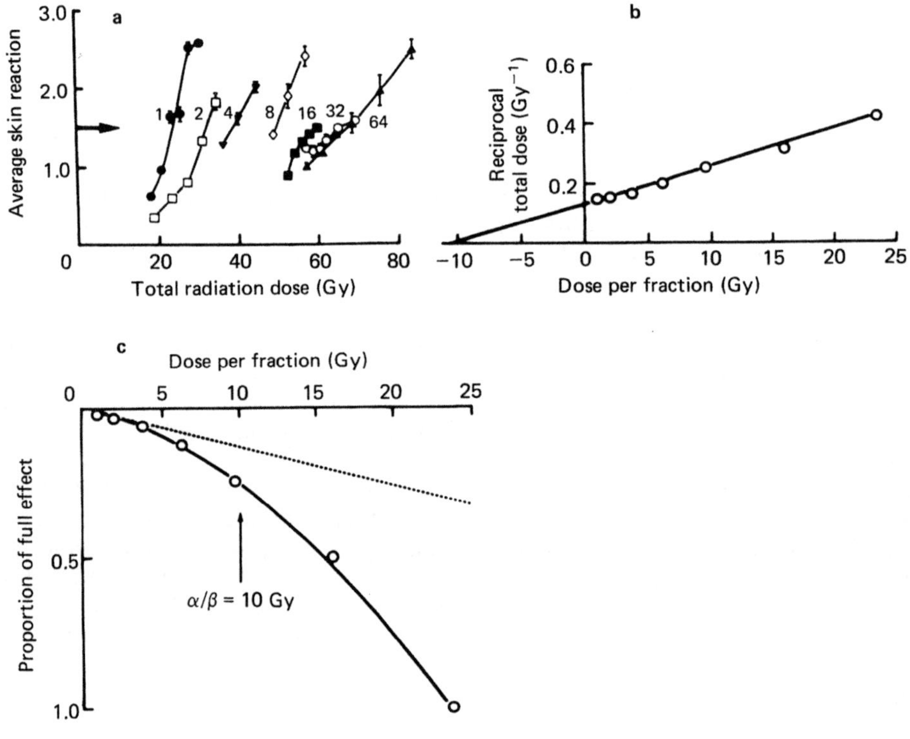

Fig. 3.4a–c. Fractionation studies in skin (from Douglas and Fowler 1976). **a** The raw dose–response curves. **b** The reciprocal dose plot shows that the data can be fitted quite well to a straight line extrapolating back to −10 Gy. **c** The underlying dose–response curve, reconstructed by assuming that the full effect is caused by a single dose, half that effect by each of four fractions, etc.

1976; Fowler 1984; or Thames & Hendry 1987; also Chaps. 6 and 7 in this volume.)

Some recent reviews of various topics of interest are listed in Table 3.2. Several clinically important findings have emerged. Early and late reacting tissues behave differently to hyper- and hypofractionation. The degree of

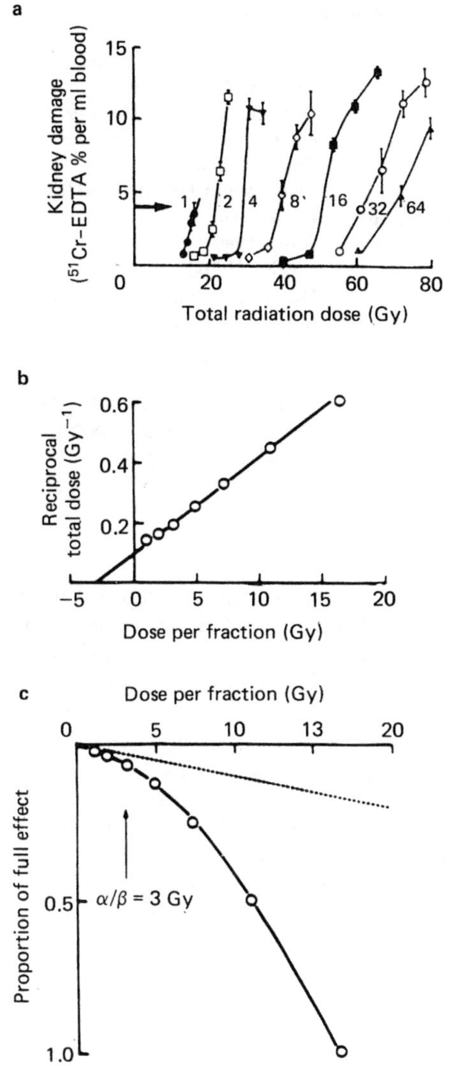

Fig. 3.5a–c. Fractionation studies in kidney (from Stewart et al. 1984). **a** The curves for 16, 32 and 64 fractions show more separation than those for skin. **b** Because the total dose varies with fractionation over the whole range, the reciprocal dose plot is steeper and extrapolates back to a value of −3 Gy. **c** The derived underlying survival curve is curvier than that shown in Fig. 3.4 for skin.

additional sparing when doses are subdivided below 2 Gy is minimal for skin and other early reacting tissues, and for most tumours. By contrast considerable extra sparing is seen in kidney, lung and other late responding tissues. The corollary is that 15–20 larger fractions of 3–4 Gy instead of 30 × 2 Gy can be disastrously effective on a slowly responding tissue like kidney, whereas 60 Gy will have almost the same effect on skin or on tumours regardless of whether it is given in 2 Gy or 3 Gy fractions.

Table 3.2. Reviews of topics relating to normal tissue response to radiation

Cellular basis of tissue response	Potten and Hendry (1985) Hendry et al. (1986)
Cell kinetics and tissue response	Denekamp and Fowler (1977) Denekamp (1984, 1986)
Fractionation	Fowler (1984) Thames and Hendry (1987)
Fast neutrons and other high linear energy transfer radiation effects	Field (1976) Raju (1980) Fowler (1981)
Hyperthermia	Dethlefson and Dewey (1982)
Combined chemotherapy and radiotherapy	Sutherland (1982)
Radiosensitizers: effects on normal tissues	Adams et al. (1978) Stewart et al. (1982) Chapman and Whitmore (1984)
Radioprotectors	Nygaard and Simic (1983) Denekamp (1984)

Acutely responding tissues can compensate for much of the damage inflicted over 6–8 weeks by increasing their production of new cells as the tissue depletion is recognized. Thus skin, intestinal and buccal mucosa can "repair" or heal much of the inflicted injury if a 2-week gap is inserted half-way through treatment. This reduces immediate morbidity, as well as increasing patient comfort and therefore compliance with the protocol. However, no such sparing occurs in the more slowly proliferating tissues which will ultimately give rise to the life-threatening morbidities months or years later. Tissue healing only commences when tissue depletion is recognized, and in kidney, lung, heart, skin, brain, bladder etc. this has a latent period of many months. Thus it is dangerous to assume that relieving early symptoms will lead to a corresponding alleviation of late injury. There is no support for this concept from the biological studies in rodents.

Effects of Associated Therapy

Radiosensitizers

Some tumour radiosensitizers, such as BUdR or IUdR, are incorporated during the S phase and therefore the degree of radiosensitization is proportional to the amount of DNA synthesized during the period of exposure to the agent. Thus rapidly proliferating tissues are more susceptible to damage than tissues with a long cell cycle time. BUdR is also a sensitizer to ultraviolet radiation (in sunlight) and therefore skin can be the limiting tissue for a reason unrelated to the X-rays. With IUdR this is less of a problem.

Other radiosensitizers are believed to have specificity for tumours because they are only effective on cells which are short of oxygen. Such cells are common in poorly perfused tumours but are rare in well-vascularized normal tissues (Adams et al. 1978; Chapman and Whitmore 1984). However, a knowledge of tissue architecture and vasculature shows that certain parts of the testis and non-vascularized cartilage are at risk from sensitizers such as misonidazole or hyperbaric oxygen. The clinical studies with hyperbaric oxygen have indeed shown some sensitization of cartilage (Henk 1981).

Radioprotectors

Thiol compounds (or tourniquet hypoxia) can be effective radioprotectors. Thiols compete with oxygen to oxidize or reduce products of radiolysis (Yuhas et al. 1980; Denekamp 1984). Thus they are most effective in conditions where the oxygen supply is just barely adequate, for then the balance for competitive reactions is most finely balanced. They are ineffective in anoxia or in very high oxygen tensions. The exact degree of oxygenation of the critical cells in a tissue will therefore determine the effectiveness of a radioprotector, which varies widely from tissue to tissue. The most promising thiol in the clinic (WR2721) is a phosphorylated compound which requires cleavage of the phosphate group before it can enter cells and before the thiol group becomes accessible. The dephosphorylation probably occurs close to the walls of the blood vessels, where phosphatases are produced. Since tumour endothelia are lacking or show low levels of alkaline phosphatase, this may explain why the compound seems to be actively excluded from solid tumours in vivo. Thus, the biochemical action of certain tissue components may influence the effectiveness of such a pro-drug that requires activation.

Hyperthermia

The effect of heat as an adjunct to radiotherapy is well documented (see Chaps. 17 and 18). Therapeutic benefit may accrue from increased tumour response. However, the side-effects of heat on normal structures are: sensitization to radiation damage and direct thermal burns. The direct effects are always seen within hours of heating, whereas the radiosensitization may remain concealed because the pattern of development of damage is the same as after radiation alone (Hill and Denekamp 1987). Thus the late morbidity is a worrying limitation of localized hyperthermia if this is combined with radiation doses that are already close to the tolerance limit.

The precise vascular pattern and the presence of reserve vessels and/or arteriovenous shunts differs from tissue to tissue. It is the ability to increase blood flow that allows normal tissues to dissipate the heat applied locally during hyperthermic therapy; poorly perfused tumours do not have this capability. Thus the tumours may actually reach a higher temperature than the normal tissue, which also gives a therapeutic advantage. If blood flow is too rapid, however, or if the fraction of cardiac output going to that tissue is too high, the net result of effective heat dissipation may be a considerable rise in total body temperature.

Summary

If the biological principles dictating the response to a modifying agent are understood it is possible to focus on the relevant characteristics of normal tissues which will limit a novel radiotherapy schedule. These may relate to cell kinetic parameters, cell–cell interdependence, biochemical peculiarities or vascular physiology. These principles have been extensively studied in small rodents and a significant body of data is now available to act as a guide to the clinician. Although direct extrapolation from mouse to man cannot be made because of differences in life span, metabolic rates and absolute organ size, nevertheless the principles can be applied, allowing new modalities to be approached with an appropriate degree of cautious optimism.

References

Adams GE, Fowler JF, Wardman P (eds) (1978) Hypoxic cell sensitizers in radiobiology and radiotherapy. Proceedings of the 8th LH Gray Conference. Br J Cancer 37 [Suppl III]

Alper T (1979) Cellular radiobiology. Cambridge University Press, Cambridge

Chapman JD, Whitmore GJ (eds) (1984) Chemical modifiers of cancer treatment. Int J Radiat Oncol Biol Phys [Special Issue] 10:1161–1812

Curtis SB (1986) Lethal and potentially lethal lesions induced by radiation – a unified repair model. Radiat Res 106:252–270

Denekamp J (1982) Cell kinetics and cancer therapy. CC Thomas, Springfield, Illinois

Denekamp J (1984) Normal tissue radioprotection by WR2721. In: Breccia A, Greenstock, CL, Tamba M (eds) Advances on oxygen radicals and radioprotectors. Lo Scarabro, Bologna, pp 153–172

Denekamp J (1986) Cell kinetics and radiation biology. Int J Radiat Biol 49:357–380

Denekamp J, Fowler JF (1977) Cell proliferation kinetics and radiation therapy. In: Becker FF (ed) Cancer: a comprehensive treatise. Plenum, New York, pp 101–128

Dethlefson LA, Dewey WC (eds) (1982) Cancer therapy by hyperthermia, drugs and radiation. National Cancer Institutes Monograph 61

Douglas BG, Fowler JF (1976) The effect of multiple small doses of X-rays on skin reactions in the mouse and a basic interpretation. Radiat Res 66:401–426

Field SB (1976) An historical survey of radiobiology and radiotherapy with fast neutrons. Curr Top Radiat Res 11:1–86

Fowler JF (1981) Nuclear particles in cancer treatment. Adam Hilger, Bristol (Medical Physics Handbooks 8)

Fowler JF (1984) What next in fractionated radiotherapy? Br J Cancer 49:285–300

Hall EJ (1978) Radiobiology for the radiologist. Harper & Row, New York

Hendry JH, Potten CS, Moore JV, Hume WJ (eds) (1986) Assays of normal tissue injury, and their cellular interpretation. Br J Cancer 53 [Suppl VII]

Henk JM (1981) Does hyperbaric oxygen have a future in radiation therapy? Int J Radiat Oncol Biol Phys 7:1125–1128

Hill SA, Denekamp J (1987) Therapeutic benefit from combined heat and radiation. In: Streffer C (ed) Hyperthermia and the therapy of malignant tumours. Springer, Berlin Heidelberg New York (Recent results in cancer research, vol 104)

Hornsey S (1973) The effectiveness of fast neutrons compared with low LET radiation on cell survival measured in the mouse jejunun. Radiat Res 55:58–68

Nygaard OF, Simic MG (eds) (1983) Radioprotectors and anticarcinogens. Academic Press, London

Potten CS, Hendry JH (eds) (1985) Manual of mammalian cell techniques. Churchill Livingstone, Edinburgh

Raju, M (1980) Heavy particle radiotherapy. Academic Press, New York
Rubin P, Casarett GW (1968) Clinical radiation pathology, vols 1 and 2. Saunders, Philadelphia
Sinclair WK (1972) Cell cycle dependence of the lethal radiation response in mammalian cells. Curr Top Radiat Res 7:264–285
Stewart FA, Denekamp J, Randhawa VS (1982) Skin sensitization by misonidazole: a demonstration of uniform mild hypoxia. Br J Cancer 48:869–877
Stewart FA, Soranson J, Alpen EL, Williams MV, Denekamp J (1984) Radiation-induced renal damage. The effects of hyperfractionation. Radiat Res 98:407–420
Sutherland RM (1982) Chemical modification: radiation and cytotoxic drugs. Int J Radiat Oncol Biol Phys 8:323–815
Thames HD, Hendry JH (eds) (1987) Fractionation in radiotherapy. Taylor & Francis, London
Withers HR, Peters LJ, Thames HD, Fletcher GH (1982) Hyperfractionation. Int J Radiat Oncol Biol Phys 8:1807–1809 (editorial)
Yuhas JM, Spellman JM, Culo F (1980) The role of WR 2721 in radiotherapy and/or chemotherapy. Cancer Clin Trials 3:211–216

4 Measurement of Human Normal Tissue and Tumour Responses

G. Ross and J. R. Yarnold

Introduction

The scarcity of quantitative measures of normal tissue damage and tumour response in patients undergoing radiotherapy is an obstacle to the clinical evaluation of new treatment strategies. Retrospective studies of complications in critical normal tissues taught important lessons in the past concerning the potential dangers of hypofractionation (Singh 1978). However, it is unethical to use serious complications as planned end-points in prospective studies. One aim of this paper is to review the desirable characteristics of clinical end-points required to compare alternative treatments employing radiotherapy, with emphasis on simple scales applied by clinicians or even the patients themselves.

Characteristics of Human Normal Tissue End-points

It is clear that the concept of a single target cell type is misleading and that a simple relationship is not expected between killing of target cells and visible damage or functional loss in a tissue or organ. There is increasing reliance on functional assays even in animal studies aimed at exploring the pathogenesis of radiation injury at a cellular level. This has important implications for the choice of end-points applied in clinical practice, where the commonest questions are concerned with measuring the effects of treatment rather than exploring mechanisms of tissue damage. It is inappropriate to develop human counterparts of laboratory end-points originally designed to help answer biological questions.

The priority is to develop clinical end-points which help to answer therapeutic questions.

Arbitrary graded scales of normal tissue injury can be powerful tools. For example, the use of a 6-point scale of acute skin reactions was applied by Fowler and his colleagues to record acute skin reactions in a series of classic pig experiments. They assigned precise quantitative values to the importance of fraction size and overall treatment time in determining the level of acute reactions to a course of fractionated radiotherapy (Fowler et al. 1963). The scale described six levels of effect, viz. no reaction, faint erythema, erythema, marked erythema, moist desquamation of less than half the irradiated area, moist desquamation of more than half the irradiated area.

Clinical scales of normal tissue damage should have the following attributes:

Clinical relevance
Reliability
Dose–effect relationship
Sensitivity
Feasibility

A fractionation trial conducted by the UK Radiotherapy and Oncology Study Group may be used to illustrate the choice of appropriate end-points. In women given breast radiotherapy after local excision for early stage breast cancer, the aim of the trial is to identify a fractionation schedule based on 13 fractions over 5 weeks which is equivalent to a daily schedule delivering 50 Gy in 25 daily fractions to breast via tangential fields. The equivalence, or lack of it, can be compared in different terms. Fibrosis might be scored subjectively using graded arbitrary scales applied by independent observers on the basis of breast palpation. Alternatively, the development of an objective score based on densitometry of computerized tomographic scans through the breasts or mammograms could be considered. These measures of fibrosis might have biological significance but their relationship to breast cosmesis may not be clear enough to be relied upon as the sole end-points. Breast cosmesis must therefore be scored in its own right and this can be approached in a number of ways using observer ratings or patient self-assessments.

Reliability is the next most important requirement of a clinical scale. Unless the scale is repeatable a high level of variance may obscure important differences in levels of normal tissue damage between patients. If objective measures are applied, repeatability is optimized by standardizing the conditions of measurement. For example, the measurement of skin erythema using reflectance spectrophotometry involves the avoidance of wide variations in ambient temperature in the clinic and the calibration of the machine using standard filters prior to each measurement. In an analogous way, the criteria used to define subjective scales must be standardized as carefully as possible. Reliability is improved by using more than one independent observer.

A clinical scale must vary in a dose-related fashion over the range of interest. For example, if telangiectasia is being used as a late end-point of skin damage, increments of radiation dose must be associated with increments of effect throughout the proposed scale.

With regard to the sensitivity of clinical scales, they need be no more precise than the clinical problem demands. In practice, 4-point scales are ample for most clinical situations and even so it is quite common to compress a scale when analysing the results. For example, in the evaluation of subcutaneous fibrosis in men previously treated by radiotherapy preceded or followed by chemotherapy for testicular teratoma, a 6-point scale was originally drawn up (none, minimal, mild, moderate, marked, severe), which was soon compressed to a 4-point scale for the purposes of examination (none, mild, moderate, severe) and further compressed to a 2-point scale for the purposes of analysis (Yarnold et al. 1983). The 6-point scale was cumbersome and failed to yield extra information concerning the effects of drug sequence on the incidence and severity of radiation-induced induration.

Finally, the utility of a clinical end-point of normal tissue damage depends partly on how easy it is to apply. There are practical and ethical constraints on tissue biopsies and sophisticated physiological tests which are time-consuming and arduous for patients to undergo. It is not possible to define rigid rules in this respect because individual patients may be willing to undergo repeated evaluations for the sake of a research project. On the other hand, end-points for application in large clinical trials must be of the simplest kind.

Application of Human Normal Tissue End-points

Graded Scales and Trial Design

The unknown relationship between arbitrary scales and target cell survival or the survival of functional units is not a serious disadvantage but has implications for how end-points are applied. Arbitrary scales are not necessarily linear and the quantitative relationship between a given level of clinical effect and the killing of putative target cells or tissue sparing units is usually unknown. It is therefore misleading to represent a dose–effect relationship in the form of a dose–response curve as shown schematically in Fig. 4.1. For example, one cannot assume that a

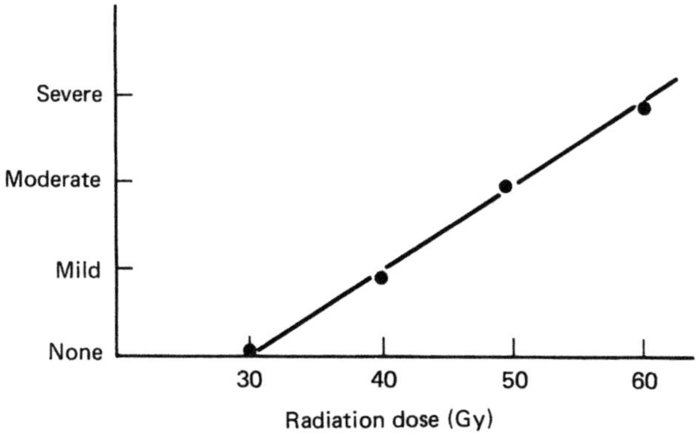

Fig. 4.1. Potentially misleading representation of a correlation between radiation dose and a graded scale of normal tissue injury.

severe effect represents three times as much damage as a mild effect. This has implications for how end-points are applied in clinical trials, where it is not possible to compare differences in the levels of normal tissue damage recorded using arbitrary scales in a quantitative manner. A good trial design for testing the anti-tumour effect of alternative radiotherapy regimes aims to compare schedules which are isoeffective in terms of normal tissue damage. The demonstration of equivalence between two schedules in terms of normal tissue injury requires that graded scales are sensitive to differences considered to be clinically important.

For example, the previously mentioned breast fractionation trial is a three-arm study (see Table 4.1). Schedule II of the trial is isoeffective with schedule I according to a modification of the Ellis formula (Orton and Ellis 1973). There is evidence that this empirical formula predicts late effects in the skin accurately when modest changes in fraction size within the range 2–4 Gy are employed, but as a precaution schedule III of the trial represents a 12% dose reduction (Turesson and Notter 1984a,b). Schedules II and III of the trial serve to test the sensitivity of the end-points used to measure normal tissue damage, viz. fibrosis based on palpation and breast cosmesis based on photographs. When 600 patients have been accrued and followed for several years it will be possible to say which of the unconventional schedules more closely resembles the conventional arm. There is an analogy here with the experimental design used in animal models to measure dose-modifying factors (such as sensitizer enhancement ratios) by means of tumour growth delay assays (see below).

Table 4.1. Randomization of the breast fractionation trial conducted by the UK Radiotherapy and Oncology Study Group

Schedule	Total dose (Gy)	No. of fractions	Time (weeks)	Dose per fraction (Gy)	Tumour dose fractionation
I	50	25	5	2	83
II	42.9	13	5	3.3	83
III	39	13	5	3.0	73

Timing and Reporting of Assessments

The timing of acute normal tissue reactions is determined mainly by the kinetics of the normal tissue, but differences between radiotherapy schedules may have an important influence on how and when comparative measurements are made. For example, when daily fractions of 2 Gy are used, mucositis in the oral cavity and pharynx becomes visible at about day 19. It then increases in severity steadily during a 6-week course of therapy and subsides over several weeks after the end of treatment. When experimental schedules of 1.6 Gy three times per day are introduced the mucositis appears earlier, reaches a higher peak (forcing a halt in treatment) and then subsides more quickly than after the end of a course of conventional treatment (Van der Schueren et al. 1983). In patients having these different forms of radiotherapy it would be inappropriate to

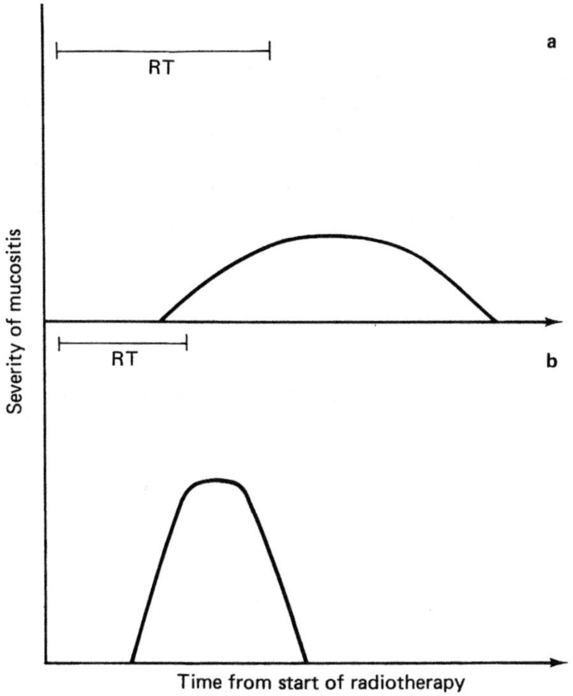

Fig. 4.2. Schematic representation of oral mucositis in patients receiving either conventional fractionation (**a**) or multiple fractions per day (**b**). RT, radiotherapy.

compare mucositis on the same day counted from the start from treatment (see Fig. 4.2). Serial measurements would be preferable but it would be misleading to compare the biological level of mucositis in terms of the area under the curves represented in Fig. 4.2. The reason for this has been mentioned, namely that the quantitative relationships between different levels of the mucositis scale are not known. If the comparison is required in strictly clinical terms, it may be better to use patient self-assessment scales to determine the tolerance to two different patterns of acute reaction.

In late responding normal tissues such as the brain, kidney, lung and vascular endothelium, the latent period between radiation exposure and expression of damage is foreshortened with increasing dose. This is a feature characteristic of hierarchical tissue organization, but even so it is usual to wait several years before scoring late tissue damage (Wheldon et al. 1982). In scoring late skin damage at least, the severity of injury at 2 years appears to predict quantitatively the levels of injury at 5 and 8 years (Turesson and Notter 1986). This is a valuable clinical observation based on careful experiment and is consistent with the expectations of a hierarchical system.

Most reported complication rates take no account of the period of follow-up and are extremely difficult to interpret. It should be routine practice to report complication rates using actuarial or life-table methods; this has obvious advantages for comparisons between studies.

Effect of Different Modalities

When radiation is combined with a different modality such as a cytotoxic agent or hyperthermia the same normal tissue scales may not be trustworthy if the pattern of damage is altered. An analogy is the use of complete response as an end-point of effect when radiotherapy is used with hyperthermia. Whereas complete response is a useful prognostic sign after radiation alone, complete tumour shrinkage is common after hyperthermia even in tumours destined to exhibit rapid regrowth.

Specific Examples: Lung and Central Nervous System

Clinical Measurement of Pulmonary Toxicity

Despite advances in dynamic assessment of respiratory function and thoracic imaging by computerized tomography (CT), there has proved to be considerable difficulty in defining relevant clinical end-points for assessing late pulmonary toxicity.

In animal models physiological "assays" such as breathing rate, carbon monoxide diffusion and changes in arterial gases which have examined dose–response relationships after whole lung irradiation have contributed to the identification and characterization of the two-phase response of normal lung to radiation. They have also helped to explain responses to fractionated treatments, with and without response modifiers (Gish et al. 1959). Many of these measurements are easy to perform in the clinic but are not sensitive enough to detect functional impairment when part of the lung is irradiated, and are not amenable to "regional" evaluation of the volume of tissue in question. Similarly, many of these investigations are subject to variation induced by intercurrent pathology as well as the presence of tumour. For example, coexisting chronic obstructive airway disease, superimposed infection distal to an obstructed bronchus, and volume loss may all influence pulmonary volume measurement and arterial gas estimations, independent of any further changes due to irradiation of normal tissue (Prato et al. 1977). Pulmonary function tests based on spirometry similarly require a higher degree of patient motivation and cooperation to give reproducible results.

Plain radiographic changes associated with acute radiation pneumonitis and late pulmonary fibrosis are well documented, with a spectrum of appearances which may or may not have symptomatic sequelae (Libshitz and Southard 1974). Arbitrary scoring by clinicians of plain chest X-ray changes have been used to investigate possible enhancement of the late pulmonary toxicity of radiation by concurrent cytotoxic drug administration. For example, in a retrospective study using an arbitrary 0–4 scale (none, mild, moderate, severe) a radiologist scored changes on plain chest X-ray following radiotherapy with or without carboplatin and etoposide (Glaholm et al. 1987). The control group receiving radiation alone were matched for dose in ret, age, sex, and field size. A significant dose–response relationship was demonstrated for radiological pneumonitis but no significant enhancement of damage was seen in the group receiving radiation plus cytotoxic therapy.

The relative insensitivity of plain chest X-ray changes to subclinical pulmonary toxicity has led to the evaluation of CT of lung in man. Changes in CT appearances following therapeutic irradiation have been described both qualitatively and quantitatively (Libshitz and Shuman 1984). Changes in pulmonary tissue density in radiation portals have been compared with results from unirradiated lung using computerized programmes. Dose–response data have been derived in patients which correlate changes in tissue density with treatment expressed as a single nominal dose value (Mah et al. 1987) (see Fig. 4.3). It is postulated that further CT studies may possibly allow derivation of a volume factor, in addition to the usual parameters of time and fraction number, to be expressed in an appropriate isoeffect formula for pulmonary toxicity.

In summary, conventional pulmonary physiological tests have proved relatively insensitive to radiation-induced pulmonary damage. Plain radiological evaluation has the advantage of low cost and availability, requiring less compliance in patients of poor performance status. Observer-based arbitrary scales of changes can produce useful data but have the disadvantage of nonlinearity of dose–response relationships. Further evaluation of CT density changes as a function of dose, fractionation and volume is required and could possibly be applied to combined modality treatment with radiation, drugs and biological response modifiers. However, availability and cost preclude its routine use.

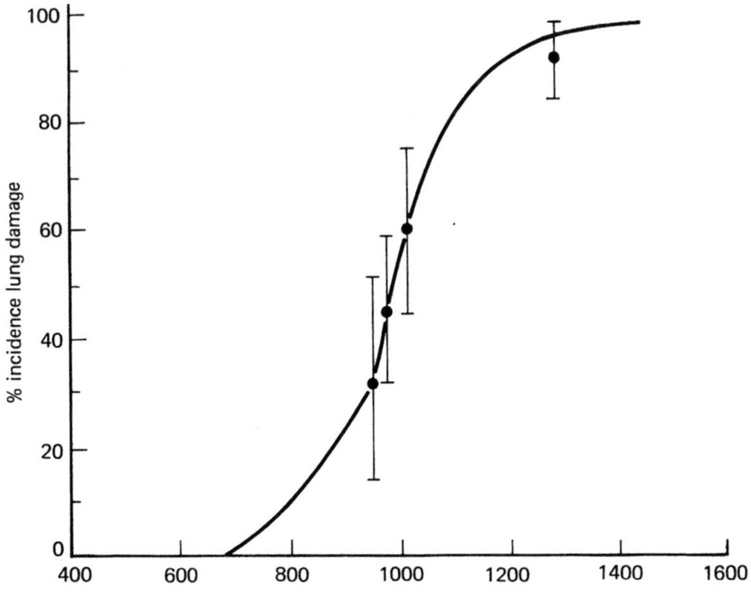

Fig. 4.3. Incidence of acute radiation-induced lung damage versus estimated single dose. (Modified from Mah et al. 1987.)

Clinical Measurement of CNS Toxicity

Over the past two decades the development of more aggressive and successful approaches to the treatment of brain tumours and extracranial malignancy in which normal brain tissue is irradiated has resulted in more critical evaluation of concepts of CNS tolerance as empirically derived (Kramer and Lee 1974; Sheline 1975). Retrospective clinical analyses have suggested modifications to the Ellis isoeffect formula, derived for skin and subcutaneous tissue but hitherto often used in CNS irradiation. The concept of the brain tolerance unit (BTU) advocated by Pezner and Archambeau (1981) is based on data collated from the retrospective analysis of clinical case records of patients identified histologically as having cerebral necrosis following a variety of tumour dose (TD) schedules given in N fractions over T days:

$$\text{TD (rad)} = \text{BTU} \times N^{0.45} \times T^{0.03}$$

The above expression yields a steeper sigmoid dose–response curve than that derived using the exponents for T and N employed in the Ellis formula and promises to be a more accurate guide to the prescribing of isoeffective schedules to the CNS (see Fig. 4.4).

Retrospective analysis of clinical complications has also contributed to our understanding of the dose–latency relationship of expression of CNS damage and has helped support hypotheses derived from animal experiments relating to specific target cells responsible for different lesions.

For example, an important contribution to our understanding of the timing of expression of human spinal cord damage has been made by Schultheiss et al. (1984). In a retrospective analysis of 77 papers reviewing over 300 cases of radiation myelopathy, they have been able to document a bimodal expression of myelopathy with peaks at 12–14 weeks and 24–28 weeks, the latent period decreasing with increasing dose. The results of this study correlate well with results of animal studies. These postulate two distinct mechanisms of late cord damage as manifest by white matter necrosis or late vascular damage, thus implicating different target cells (Van der Kogel 1986).

A more precise knowledge of the dose–latency relationship of damage may alter clinical practice in several instances. For example, a clinician may elect to exceed the tolerance of brain or spinal cord if palliation of symptoms can usefully be achieved and life expectancy is short relative to the expected latent period for expression of damage. In the analysis of clinical toxicity data allowance must be made for the loss of patients whose tolerance was exceeded but who did not live to express the injury.

The improvement in long-term survival of patients with childhood medulloblastoma, and the incorporation of prophylactic cranial irradiation as part of the aggressive management of certain acute leukaemias (ALL), has led to concern regarding the means of documenting adequately late changes in higher mental function. Early investigators confined observations to gross functional categories, that is active life, partial disability, total disability (Bloom et al. 1969). There are now a number of studies following successful treatment of medulloblastoma and ALL by radiotherapy in combination with cytotoxic therapy. These report late toxicity effects ranging from no definite neuropsychiatric sequelae to definite late intellectual impairment (Obetz et al. 1979; Rowland et al. 1984).

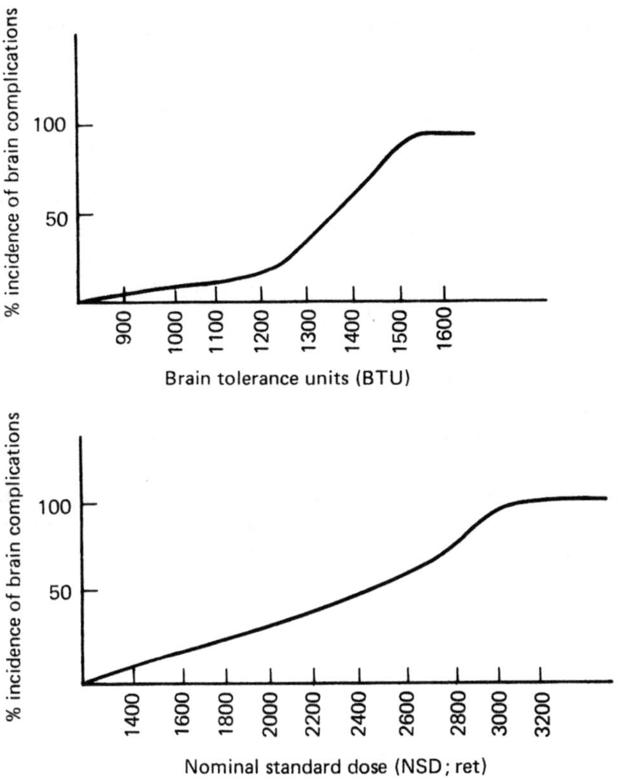

Fig. 4.4. Dose–response curves for brain complications as a function of total dose based on alternative exponents to N and T (see text). (Modified from Pezner and Archambeau 1981.)

Even in asymptomatic children serial CT scans of the brain have demonstrated varying degrees of cortical atrophy (Ricardi et al. 1985). The derivation of useful dose–response or dose–latency data has been precluded by the multivariant nature of the data: small sample sizes, variable patient and drug treatment characteristics, varying time elapsed between treatment and evaluation, variation in the instruments used for psychometric assessments and the difficulty of identifying satisfactory control groups.

Measurement of Tumour Response

Local Control

Local control refers to persistent absence of detectable tumour within an irradiated volume. Although local control is an unambiguous clinical end-point its relationship to clonogenic cell killing is not entirely clear. It is possible that host mechanisms contribute to the destruction of limited numbers of clonogenic cells remaining after curative treatment, although there is no evidence that this occurs in humans. Personal cure within the irradiated volume may not equate to biological cure if the patient dies of metastases prior to local regrowth or if surviving clonogenic cells are entrapped for many years in dense radiation-induced fibrosis.

Local control is used as the main end-point of effect in clinical trials of biological response modifiers such as hypoxic sensitizers. In the past many trials were undertaken in tumours where the expected increase in biological effect by the addition of a chemical response modifier could not be expected to result in a measurable increase in local control (Dische and Saunders 1980). An example is the evaluation of misonidazole in gliomas, where a 20% increase in radiation effect (sensitizer enhancement ratio 1.2) could not reasonably be expected to improve local control. A suitable tumour type for clinical trials of chemical modifiers or unorthodox fractionation schedules is characterized by a steep dose–response relationship above the current limits of total radiation dose. In this case small increments in biological effect will lead to a sizable improvement in local control rate. A recent review of clinical dose–response curves suggests that suitable sites include carcinomas of the bladder, cervix, and head and neck region (Williams et al. 1984).

Volume Growth Delay

Local control is a late measure of tumour response and as such is not an ideal experimental end-point. Volume growth delay is a well-established end-point in animal systems and has been introduced into the clinical context (Thomlinson et al. 1976; Ash et al. 1979; Yarnold et al. 1986). Calipers or ultrasound have been used to make serial volume measurements of multiple cutaneous metastases in individual patients and the time taken for shrinkage and regrowth to pretreatment size has been shown to be a sensitive function of dose. This system has been successfully applied to a few patients treated with small doses of radiotherapy with or without misonidazole.

Although access to and measurement of superficial tumours can be difficult, the real limitation of this technique appears to be the small number of evaluable patients. For example, in a 14-month period 42 patients with two or more superficial metastases were entered into specific protocols which used growth delay as an end-point, but only 4 patients remained fit to complete the protocols (Yarnold et al. 1986). Sixteen out of 191 (8.5%) nodules were measured to the point at which they regained their pretreatment size, reflecting a large investment in time and effort for a modest result.

Tumour Shrinkage

Measurements of tumour shrinkage are only worth while if they serve to discriminate between curable and incurable tumours early enough in the course of treatment. Then alternative or additional treatments may be delivered

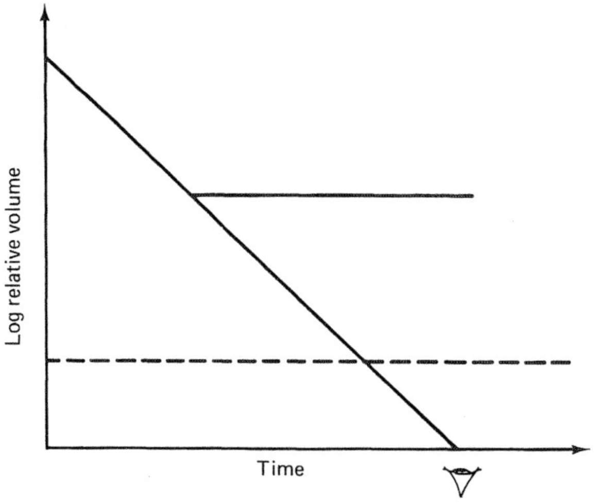

Fig. 4.5. Schematic representation of two tumours of equal initial size regressing at the same rate with one complete and one partial response. The *horizontal dashed line* shows the arbitrary limit of clinical detection.

successfully – for example salvage surgery. For several decades rapid shrinkage of a tumour during radiotherapy has maintained clinical optimism regarding its radiocurability, yet the relationship between radioresponsiveness and radiocurability is far from clear. Although the most radiocurable tumours such as lymphomas and seminoma usually shrink faster than carcinomas, within a particular tumour type the correlation is much less obvious. There appears to be no correlation between the radioresponsiveness and radiocurability of breast cancer, whereas significant correlations have been reported in some studies of head and neck cancer but not in others (Suit et al. 1965; Suit and Walker 1980; Thomlinson et al. 1976; Bartelink 1983). Radioresponsiveness can be regarded as a biological characteristic of tumours in which the rate of volume loss is determined by the rate at which dead cells and stroma are removed rather than by the rate at which cells are killed. This appears to be the case for breast cancer.

Part of the confusion may reflect the timing of clinical assessments in some reports which document the full extent of shrinkage rather than shrinkage rate. This point is represented schematically in Fig. 4.5, which depicts two tumours shrinking at the same relative rate after the start of treatment, one of which achieves complete response and the other a partial response. The observer monitoring response several weeks after the start of treatment here measures the extent of regression rather than the rate of regression. As such, it is not surprising if assessments at this time are good prognostic indicators.

Differences in tumour size at the start of treatment introduce a potential source of bias when considering the prognostic significance of shrinkage rates. This is depicted schematically in Fig. 4.6, which represents the shrinkage curves of two tumours differing in initial size. Small tumours disappear sooner than large tumours without necessarily shrinking any faster. Tumours that disappear

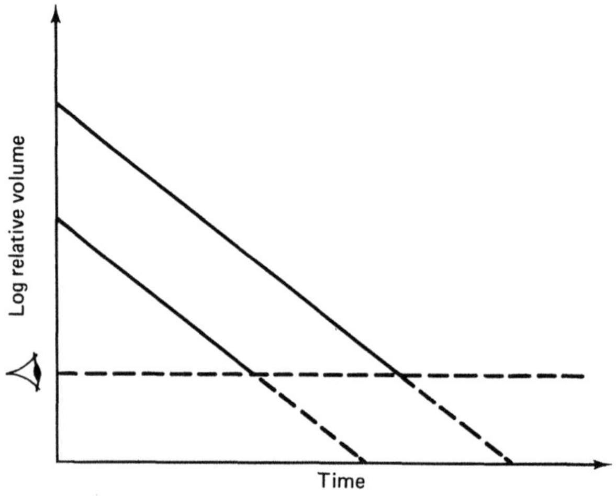

Fig. 4.6. Schematic representation of two tumours of differing initial size regressing at the same relative rate to complete response.

first may be more radiocurable, but this characteristic may be more strongly related to tumour size than to radioresponsiveness *per se*.

Biochemical Assays of Response

Despite much research directed at identifying tumour-specific products, only human chorionic gonadotrophin, alphafetoprotein and myeloma protein are of clear clinical value. Other products such as carcinoembryonic antigen, neurogenic amines and serotonin metabolites remain of experimental interest only. As yet the potential clinical applications of isotopically labelled monoclonal antibodies to tumour-specific products or antigens remain under investigation.

Histological Assessments

Morphological assessments of biopsy material taken during and after radiotherapy for carcinoma of the cervix indicate a correlation between the number of viable cells remaining and subsequent control of the tumour (Glucksmann and Spear 1945; Glucksmann 1974). However, it appears that the correlation may not be as strong as that obtained by monitoring the extent of regression at the end of a course of treatment (Dische and Saunders 1980).

Surgical resection following curative radiotherapy has yielded reliable measurements of tumour response. For example, trials of preoperative radiotherapy in non-small-cell lung cancer have shown that fractionated treatment to total doses of 40–45 Gy sterilizes 20% of tumours (Katz 1983). Similar information can be gained at autopsy, but in practice post-mortem rates are low unless special efforts are made to increase them (Saunders et al. 1982).

Excisional Assays

The low plating efficiency of many human tumours discourages clinical experiments which attempt to perform a clonogenic assay on single cell suspensions derived from different superficial nodules irradiated under different conditions. Recent experience with short-term cultures which do not measure clonogenic ability but quantitate cell viability by uptake of crystal violet offers scope for interesting clinical applications (Peters et al. 1986). Although clear dose–response relationships exist in short-term assays of human tumour cell lines it is not yet clear whether they will predict for local control in vivo.

Measurement of Patient Response

The value of patient self-assessments using validated questionnaires as quantitative measures of early and late normal tissue damage has already been described. In the large proportion of patients who cannot reasonably expect a cure from treatment, the rationale for improvements in current treatment must be to improve the quality of remaining life. Precise measures of tumour response and normal tissue reaction do not necessarily predict the degree of patient benefit. Global measures such as the Karnofsky performance status or responses to a question such as "How do you rate your quality of life today?" may correlate with more detailed assessments but fail to provide any insight into why a patient is more or less satisfied with the outcome of treatment.

The Medical Research Council has developed a General Health Diary Card which is completed by patients at home on a daily basis. It is used as one of the end-points in a prospective randomized trial which is comparing the therapeutic response of 30 Gy in 10 daily fractions with 2 fractions of 8.6 Gy spaced 1 week apart for the palliation of patients with inoperable non-small-cell lung cancer. This questionnaire asks patients to rate six items on a 4- or 5-point scale, viz. nausea, vomiting, soreness in swallowing, mood, activity and overall condition.

The EORTC Lung Cancer Cooperative Group has developed and extensively tested more detailed questionnaires with up to 30 items which can be answered in less than 10 minutes and which tap several domains including physical performance, symptoms of lung cancer, side-effects of treatment, anxiety, depression, social interactions and functional status (Aaronson et al. 1986). This instrument is being used as one end-point in a current randomized trial of cisplatin as a radiosensitizer in patients with inoperable non-small-cell lung cancer.

Concluding Remarks

A wide range of simple quantitative measures of normal tissue injury should be part of the repertoire of every radiotherapist and oncologist involved in clinical protocols. It is hoped that reliable predictive assays of both normal tissue injury and tumour response will greatly improve the effectiveness of clinical radiobiology in radiotherapy practice.

References

Aaronson NK, Bakker W, Stewart AL et al. (1986) A multi-dimensional approach to the measurement of quality of life in lung cancer clinical trials: a joint project of the EORTC Lung Cancer Cooperative Group and Study Group on quality of life. In: Aaronson NK, Beckmann J, Bernheim J, Zittoun R (eds) Quality of life in cancer. Raven Press, New York, pp 63–82 (EORTC monograph series)

Ash D, Peckham MJ, Steel GG (1979) The quantitative response of human tumours to radiation and misonidazole. Br J Cancer 40:883–889

Bartelink H (1983) Prognostic value of the regression rate of neck node metastases during radiotherapy. Int J Radiat Oncol Biol Phys 9:993–996

Bloom HJG, Wallace ENK, Henk JM (1969) The treatment and prognosis of medulloblastoma in children: a study of 82 verified cases. Am J Roentgenol 105:43–62

Dische S, Saunders MI (1980) Tumour regression and prognosis: a clinical study. Br J Cancer 41 [Suppl IV]:11–13

Fowler JF, Morgan RL, Silvester JA, Bewley DK, Turner BA (1963) Experiments with fractionated X-ray treatment of the skin of pigs. Br J Radiol 36:188–196

Gish JR, Coates EO, Dussault AB, Doub H (1959) Pulmonary radiation reaction: a vital capacity and time–dose study. Radiology 73:679–683

Glaholm J, Repetto JR, Yarnold JR, Smith IE, Magrini S, Cherryman G. (1987) Carboplatin (JM8), etoposide (VP16) and thoracic irradiation for small cell lung cancer (SCLC): an evaluation of lung toxicity (Submitted)

Glucksmann A (1974) Histological features in the local radiocurability of carcinomas. In: Friedman M (ed) The biological and clinical basis of radiosensitivity. CC Thomas, Springfield, Illinois, pp 203–218

Glucksmann A, Spear FG (1945) The qualitative and quantitative histological examination of biopsy material from patients treated by radiation for carcinoma of the cervix. Br J Radiol 18:313

Katz HR (1983) The effect of resection on local failure in irradiated non-oat cell carcinoma of the lung. Int J Radiat Oncol Biol Phys 9:1793–1805

Kramer S, Lee KF (1974) Complications of radiation therapy: the central nervous system. Semin Roentgenol 9:75–83

Libshitz HI, Shuman LS (1984) Radiation induced pulmonary change: CT findings. J Comput Assist Tomogr 8:15–19

Libshitz HI, Southard ME (1974) Complications of radiation therapy: the thorax. Semin Roentgenol 9:41

Mah K, Van Dyk J, Keane T, Poon PY (1987) Acute radiation-induced pulmonary damage: a clinical study on the response to fractionated radiation therapy. Int J Radiat Oncol Biol Phys 13: 179–188

Obetz SW, Smithson WA, Groover RV et al. (1979) Neuropsychologic follow-up study of children with acute lymphocyte leukaemia: a preliminary report. Am J Paediatr Hematol Oncol 1:207–213

Orton CG, Ellis F (1973) A simplification in the use of NSD concept in practical radiotherapy. Brit J Radiol 46:529–537

Peters LJ, Brock WA, Johnson T, Meyn RE, Tofilon PJ, Milas L (1986) Potential methods for predicting tumour radiocurability. Int J Radiat Oncol Biol Phys 12:459–467

Pezner RD, Archambeau JO (1981) Brain tolerance unit: a method to estimate risk of radiation brain injury for various dose schedules. Int J Radiat Oncol Biol Phys 3:397–402

Prato FS, Kurdyak R, Saibil E, Rider W, Aspin N (1977) Regional and total body lung function in patients following pulmonary irradiation. Invest Radiol 12:224–237

Riccardi R, Brouwers P, Di Chiro G, Poplack DG (1985) Abnormal computed tomography brain scans in children with acute lymphoblastic leukaemia: serial long-term follow-up. J Clin Oncol 3:12–18

Rowland JH, Glidewell OJ, Sibley RF et al. (1984) Effects of different forms of central nervous system prophylaxis on neuropsychologic function in childhood leukaemias. J Clin Oncol 2:1327–1335

Saunders MI, Anderson P, Dische S, Craig Martin WM (1982) A controlled clinical trial of misonidazole in the radiotherapy of patients with carcinoma of the bronchus. Int J Radiat Oncol Biol Phys 8:347–350

Schultheiss TE, Higgins EM, El-Mahdi AM (1984) The latent period in clinical radiation myelopathy. Int J Radiat Oncol Biol Phys 10:1109–1115

Sheline GE (1975) Radiation therapy of primary tumours. Semin Oncol 2:29–42

Singh K (1978) Two regimes with the same TDF but differing morbidity used in the treatment of stage III carcinoma of the cervix. Br J Radiol 51:357–362

Suit HD, Walker AM (1980) Assessment of the response of tumours to radiation: clinical and experimental studies. Br J Cancer 41 [Suppl IV]:1–10

Suit H, Lindberg R, Fletcher GH (1965) Prognostic significance of extent of tumour regression at completion of radiation therapy. Radiology 84:1100–1107

Thomlinson RH (1982) Measurement and management of carcinoma of the breast. Clin Radiol 33:481–493

Thomlinson RH, Dische S, Gray AJ, Errington LM (1976) Clinical testing of the radiosensitiser Ro–07–0582. 3. Response of tumours. Clin Radiol 27:167

Turesson I, Notter G (1984a) The influence of fraction size in radiotherapy on the late normal tissue reaction. 2. Comparison of the effects of daily and twice a week fractionation on human skin. Int J Radiat Oncol Biol Phys 10:599–606

Turesson I, Notter G (1984b) The influence of fraction size in radiotherapy on the late normal tissue reaction. 1. Comparison of the effects of daily and once a week fractionation on human skin. Int J Radiat Oncol Biol Phys 10:593–598

Turesson I, Notter G (1986) The predictive value of skin telangiectasia for late radiation effects in different normal tissues. Int J Radiat Oncol Biol Phys 12:603–609

Van der Kogel AJ (1986) Radiation-induced in the CNS: an interpretation of target cell responses. Br J Cancer 53 [Suppl VII]:207–217

Van der Schueren E, Van den Bogaert W, Kian Ang K (1983) Radiotherapy with multiple fractions per day. In: Steel GG, Adams GE, Peckham MJ (eds) The biological basis of radiotherapy. Elsevier, Amsterdam, pp 195–210

Van Dyk J, Hill RP (1983) Post-irradiation lung density changes measured by computerised tomography. Int J Radiat Oncol Biol Phys 9:847–852

Wheldon TE, Michalowski AS, Kirk J (1982) The effect of irradiation on function in self-renewing normal tissues with differing proliferative organisation. Br J Radiol 55:759–766

Williams MV, Denekamp J, Fowler JF (1984) Dose–response relationships for human tumours: implications for clinical trials of dose modifying agents. Int J Radiat Oncol Biol Phys 10:1703–1707

Yarnold JR, Horwich A, Duchesne G, Westbrook K, Gibbs JE, Peckham MJ (1983) Chemotherapy and radiotherapy for advanced testicular non-seminoma. 1. The influence of sequence and timing of drugs and radiation on the appearance of normal tissue damage. Radiother Oncol 1:91–99

Yarnold JR, Bamber, JC, Gibbs J (1986) Tumour growth delay as a clinical endpoint for the measurement of radiation response. Radiother Oncol 5:207–214

5 Late Effects of Radiation in Man

R. J. Berry

Introduction

Other contributors to this volume have dealt in detail with the biological basis of radiation damage to normal tissues unavoidably included in the treatment volume. However, it is perhaps salutary to remember that the greatest radiotherapeutic advance of the last 30 or so years has been that the *physical* dose-sparing of the skin by megavoltage X-rays has dramatically reduced the quite visible late damage which all radically treated patients suffered in the past following treatment with orthovoltage X-rays. Another tremendous advance brought about by the physical characteristics of more penetrating megavoltage beams and by the recent step-change in our capabilities for tumour and normal tissue imaging brought about by computerized tomography, ultrasound and magnetic resonance imaging is the ability to limit more precisely the high-dose volume in which the majority of late effects are seen to the immediate environs of the tumour.

There is a central mechanism underlying the dose-related limiting effects on normal tissues and this is in large part related to damage to blood vessels (Hopewell 1986; Hopewell et al. 1986). Early damage to the endothelial lining of small vessels is associated with transient vascular leakage, but the major damage is observed only months or years later when the endothelial/intimal cells of the blood vessels attempt to replace themselves by division. Clonal proliferation of the surviving endothelial cells causes small vessel obstruction. The disturbance of blood flow downstream of such obstruction leads to the body's normal response to hypoxia, that is the deposition of fibrin and the replacement of more rapidly metabolizing tissues by fibrous connective tissue with the laying down of extensive collagen. The process is unmistakably related to the magnitude of the radiation dose delivered and, as other contributors to this volume have noted, its severity can be affected by the dose/time pattern in which radiation is delivered as well as by the specific body site and the volume of tissue irradiated.

However, there is another late sequel of radiotherapy which, although rare, is not a zero risk and is related to the volume of tissue irradiated and the dose received in the low-dose regions *outside* the target volume. Radiation injury is classified as stochastic or non-stochastic. Non-stochastic injuries are those for which the severity of the injury is proportional to the radiation dose, but *only* when a threshold sufficient to allow the injury to be seen at all has been exceeded. The late effects of radiation observed in the high-dose volume, such as fibrosis or late tissue necrosis, and the late effects seen in the past on the skin, such as dyspigmentation and telangiectasis, are non-stochastic effects. Stochastic effects, on the contrary, are those for which the severity of the damage is independent of the radiation dose. Only the risk of the untoward event taking place is related to the radiation dose. For man, the major stochastic risk of radiation exposure is the subsequent development of malignant disease.

Radiation Carcinogenesis

Radiation-induced cancer was first observed in the early radiologists and their patients. In more recent years the largest accumulation of evidence of the risk of radiation induction of a new malignancy has come from the treatment of non-malignant conditions.

It was noted by Court Brown and Doll in the late 1950s that young men treated with radiotherapy for ankylosing spondylitis had an elevated risk of developing leukaemia, and further follow-up of these patients has shown significantly increased risks of developing more common tumours such as those of the bronchus and large bowel (Court Brown and Doll 1957; Darby et al. 1987). Comparable long-term follow-up studies of ankylosing spondylitics not treated by radiotherapy have found that the significantly elevated risk of dying from non-malignant causes, which patients with the disease show when compared with age- and sex-matched controls, is related to their disease; only the increased risk of the development of cancer is related to radiation exposure (Smith et al. 1977). Attempts at developing a dose–response relationship based on average doses across organs in which tumours have subsequently developed have been less than completely successful. Hence, for planning purposes (and in discussion with patients of the potential risks and benefits of treatment) an assumption has been made that there is no threshold below which radiation is ineffective in increasing the risk of developing malignant disease. Also it is accepted that the risk increases linearly with radiation dose. Clearly, this is an over-simplified model as relatively large radiation doses are equally capable of depriving cells of reproductive capability and hence of blocking the ability of a transformed cell to proliferate as a clone and subsequently express itself as a cancer. The cumulative human experience of cancer induction following therapeutic radiation exposure is reviewed in the 1986 report of the United Nations Scientific Committee on the Effects of Atomic Radiation (UNSCEAR 1986).

In addition to the complications of the expected radiation dose response for induction of cancer, a further uncertainty arises as to whether the risk of development of malignancy following exposure is simply an *absolute* increase which is proportional to the radiation dose, or a *relative* increase which is proportional to the underlying spontaneous risk of developing a malignancy in that site. Although these two models predict widely differing numbers of patients who will develop radiation-induced malignancy, there are far smaller differences in the years of life lost from such malignant disease predicted by the two different models, since most spontaneous cancers develop in the sixth, seventh and eighth decades of life (Darby and Muirhead 1987). Evidence from patients irradiated for non-malignant conditions such as chronic mastitis and from women who received multiple fluoroscopic examinations to monitor artificial pneumothorax in the treatment of tuberculosis, have shown that the induction of solid tumours may take 20, 30, 40 or more years (UNSCEAR 1986) and hence the cumulative risk of radiation-induced second malignancy is a function not only of the radiation dose but also of the age of the irradiated individual or the age distribution of irradiated populations in which the risk will be expressed (see Fig. 5.1)

It is important not to overstate the risk, however, as there are populations in which no evidence of excessive cancer has been seen following widespread (but low-level) radiation exposure. An example is the use of radioactive iodine in the treatment of goitre; in these patients no convincing evidence has yet been adduced of an increased risk of subsequent malignancy. However, amongst the

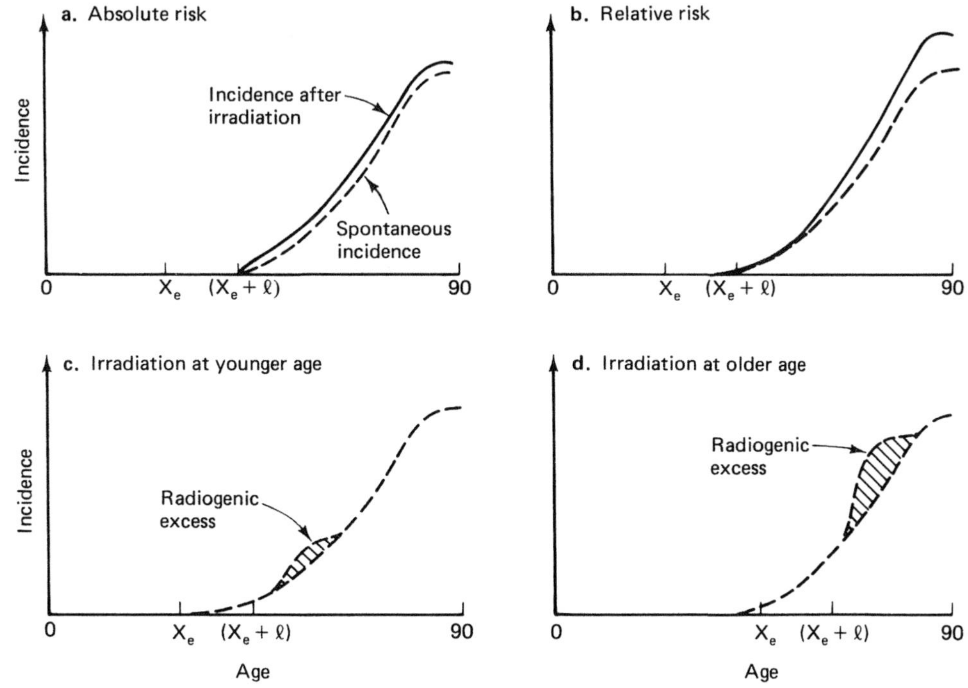

Fig. 5.1a–d. Incidence of radiation-induced cancer superimposed on incidence of spontaneous cancer by age. **a** and **b** Absolute and relative risk models. **c** and **d** Comparison of two age-at-exposure groups assuming a limited expression time, relative risk model. X_e, age at exposure; ℓ, minimum latent period. (After UNSCEAR 1986.)

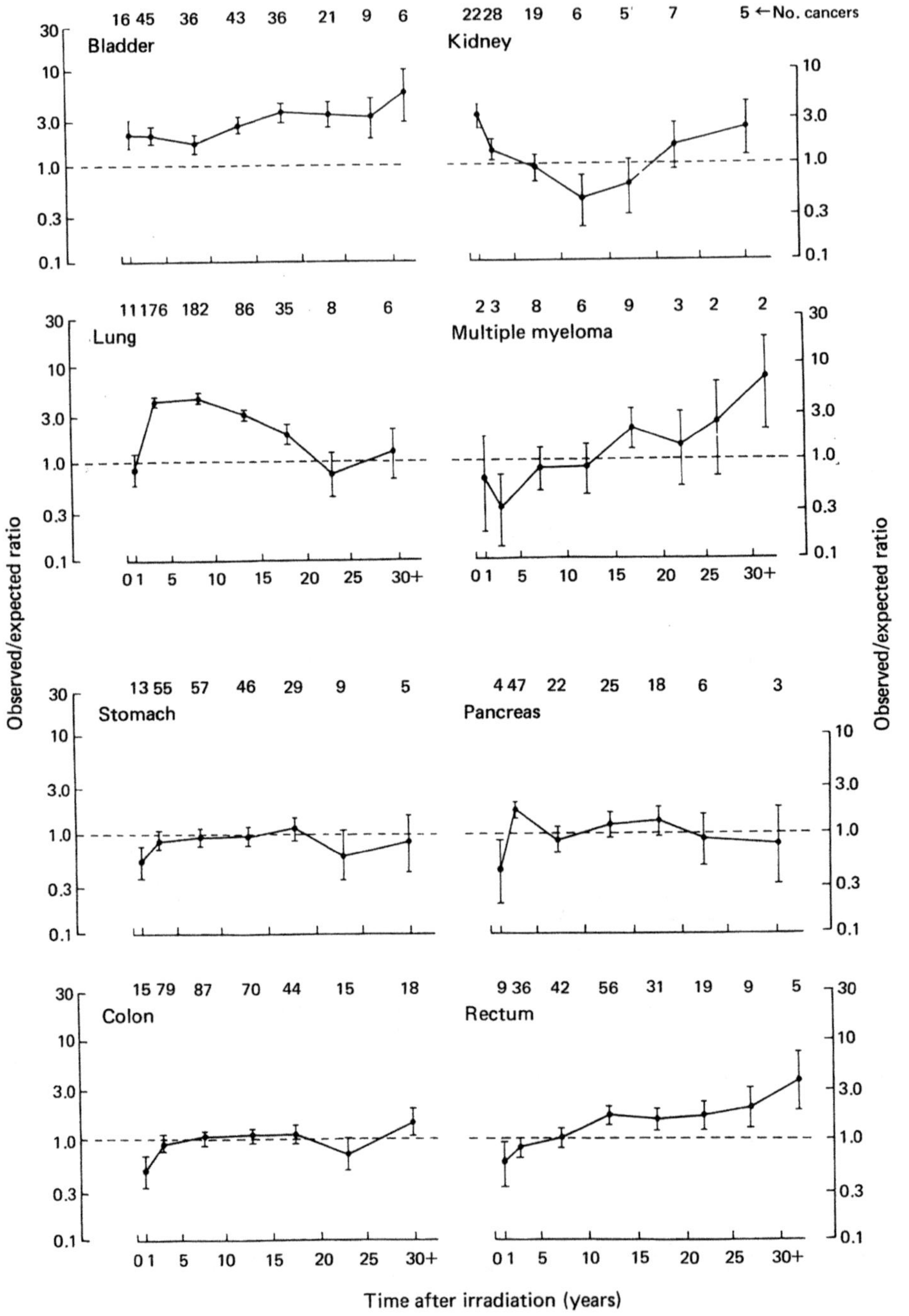

Fig. 5.2. Observed/expected ratios for selected second cancers, by time since diagnosis of cervical cancer, for women treated with radical radiotherapy. (After Boice et al. 1985.)

Fig. 5.2 (cont.)

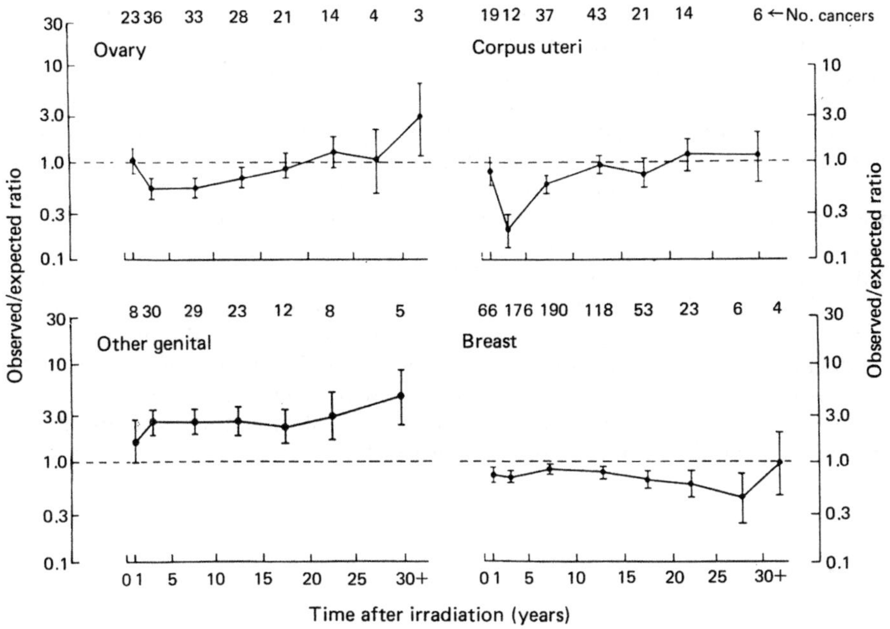

Marshallese Island population exposed accidentally to radioactive iodine in the fall-out from nuclear weapons testing, a higher subsequent risk of the development of both benign and malignant thyroid tumours has been shown (Conard et al. 1980).

Do patients treated to the limit of normal tissue tolerance in the attempt to cure malignant disease develop second cancers? Overall radiotherapeutic experience suggests that this is exceedingly rare. The largest single series of observations arises from an international collaboration amongst 15 cancer registries in eight countries which recorded second cancers following radiation treatment for carcinoma of the cervix (Boice et al. 1985). Of 182 040 women treated for cervical cancer there was only a 9% excess of second cancers (5146 observed vs. 4736 expected) which occurred more than 1 year after treatment. In 82 616 women treated by radical radiotherapy in this sample, at most about 162 of 3324 second cancers (approximately 5%) could be attributed to the radiation exposure. Significantly, only a slight excess of acute and non-lymphocytic leukaemia was found amongst the women who received radiotherapy for cervical cancer (relative risk 1.3:1); substantially fewer cases were observed than were expected on the basis of current radiation risk estimates (ICRP 1977). It was felt that this might be related to the relatively high doses to the pelvic bone marrow resulting in local ablation of stem cell proliferative capability, and the low doses received by other areas of bone marrow. The relative risk of developing cancer in organs close to the cervix that have received high radiation doses (most notably the bladder, rectum, corpus uteri, ovary) and for developing multiple myeloma increased with time after treatment (see Fig. 5.2).

No similar increase with time was seen in the 99 424 women not treated with radiation. Women who were under 30 at the time they were irradiated had the greatest risk of developing a second cancer as the expression period for radiation-induced solid tumours appeared to continue to the end of life. Transient decreases in incidence of some tumours such as those of the ovary and corpus uteri for periods of 10 or more years may reflect direct (and accidental) killing of early tumour clones already arising spontaneously, while the overall cancer incidence was affected by altered hormonal factors consequent on ovarian irradiation, so that these patients had a deficit of breast cancers compared with the unirradiated population for more than 30 subsequent years.

Thus, radiation-induced second tumours after radiotherapy are a rare but well-documented risk. It is the duty of the radiotherapist to minimize such risks by careful treatment planning. Just as it is important to minimize the treatment volume which receives a high radiation dose, in order to limit the non-stochastic late effects of radiation, so also it is necessary to be aware of and limit the volume of the body outside the target volume which receives doses in the range of a few gray. For it is in these areas that the risk of late stochastic effects, the induction of second tumours, will be expressed.

References

Boice JD, Day NE, Andersen A et al. (1985) Second cancers following radiation treatment for cervical cancer. An international collaboration among cancer registries. J Natl Cancer Inst 74:955–975

Conard RA, Paglia DE, Larsen PR et al. (1980) Review of medical findings in a Marshallese population twenty-six years after accidental exposure to radioactive fallout. BNL–51261, Brookhaven National Laboratory, Upton New York

Court Brown WM, Doll R (1957) Leukaemia and aplastic anaemia in patients irradiated for ankylosing spondylitis. Her Majesty's Stationery Office, London

Darby SC, Muirhead CR (1987) Modelling the relative and absolute risks of radiation-induced cancers. J R Stat Soc Series A 150 (2):83–118

Darby SC, Doll R, Gill SK, Smith PG (1987) Long term mortality after a single treatment course with X-rays in patients treated for ankylosing spondylitis. Br J Cancer 55:179–190

Hopewell JW (1986) Mechanisms of the action of radiation on skin and underlying tissues. Br J Radiol [Suppl] 19:39–47

Hopewell JW, Campling D, Calvo W et al. (1986) Vascular irradiation damage: its cellular basis and likely consequences. Br J Cancer [Suppl VII] 53:181–191

International Commission on Radiological Protection (ICRP) (1977) Recommendations of the international commission on radiological protection. Annals of the ICRP 1: No. 3 (ICRP publication 26)

Smith PG, Doll R, Radford EP (1977) Cancer mortality among patients with ankylosing spondylitis not given X-ray therapy. Br J Radiol 50:728–734

United Nations Scientific Committee on the Effects of Atomic Radiation (UNSCEAR) (1986) Genetic and somatic effects of ionizing radiation. Annex B, dose–response relationships for radiation-induced cancer. United Nations, New York, pp 165–262

6 Principles of Fractionation in Radiotherapy

J. F. Fowler

Introduction

"Differing fractionation schedules can be considered as a modality in the same sense as high LET radiation, sensitizers or biological modifiers: they might increase the probability of tumour control, decrease effects on normal tissues, or combine optimally with other modalities. In addition, fractionation is the backbone of all research in radiation therapy, since some time–dose regimen must be chosen any time another modality is superimposed". (Cox 1985)

The fractionation factors which the radiotherapist chooses are: (1) dose per fraction, (2) number of fractions (and consequently total dose), (3) interval between fractions, and (4) overall time. Let us consider how the choice of total dose depends upon these factors.

Overall Time

The total dose has to be increased if the overall time is increased for two reasons. The first is that if smaller doses per fraction are used, each is then less effective than larger fractions. The second reason is to compensate for proliferation in relevant tissues, namely tumours and early responding tissues. Little or no proliferation occurs in late responding tissues, so total dose should not be altered with changes in overall time when late reactions are dose-limiting.

For early reacting tissues, however, there is a period of 2–4 weeks after starting radiotherapy before compensatory proliferation begins, in mucosa and skin respectively. For longer overall times the total dose necessary to produce an acute reaction rises rapidly. It would continue to rise for many weeks, because compensatory proliferation in skin and mucosa can speed up almost to keep pace with the daily killing of cells. This is why total dose is less critical for early reactions or for tumour control than for late reactions. The time profile of extra total dose to compensate for proliferation is not at all like those represented by the nominal standard dose (NSD), time dose factor (TDF) or cumulative radiation effect (CRE) because it is the sigmoid shape just described, and not an initial rapid effect that gradually decreases.

As long as the early reacting tissues are proliferating faster than the tumour, prolongation of treatment time makes sense – with one vital proviso: the escalating total dose soon becomes limited by late reactions, because they are *not* spared by prolongation. If, however, the tumour cells are proliferating faster than cells in early responding tissues, prolongation is obviously not useful and accelerated fractionation should be considered (Trott and Kummermehr 1985; Fowler 1986a). Since human tumours have wide ranges of proliferation rates, no universal plan can apply to the use of short or accelerated schedules; indeed indiscriminate application could lead to disappointment. Methods of measuring proliferation rates in individual tumours are necessary. Fortunately rapid methods are now available and are being evaluated (Wilson et al. 1985). It has been suggested that if the potential doubling time, T_{pot}, is shorter than about 5 days in the unirradiated tumour, then that tumour should be considered a candidate for accelerated fractionation (Thames et al. 1982).

Some enterprising methods of shortening the overall treatment time are now being developed. Svoboda (1978) has been a pioneer in 10–15-day schedules. The 36 fraction/12-day schedule of Saunders and Dische (1986) keeps irradiating the tumours 3 times a day without a weekend gap, to total doses of 54 Gy. A minimum interval of 6 hours between fractions is maintained, which is important especially for avoiding late reactions. Peters and Ang (1986) described a method of concomitant boost, where the last 2 weeks' worth of smaller-volume fractions are added into the fourth and fifth weeks of treatment, making two fractions a day in these 2 weeks. They claim that full doses of 65–70 Gy can be administered in an overall time of 5 weeks instead of 7 weeks.

Whether 70 Gy in 5 weeks is better for local control than 54 Gy in 12 days obviously depends on how rapidly a given tumour is proliferating during treatment. There is not much doubt that 54 Gy will cause less late damage than 70 Gy (overall time not being a major factor here), unless the 54-Gy schedule uses larger doses per fraction than the other. In the examples quoted here, Saunders and Dische (1986) use small doses per fraction of 1.4 or 1.5 Gy in order to minimize the late reactions.

Other schedules have used multiple fractions per day (Wang 1985; van der Schueren et al. 1985) but gaps in treatment meant that the overall times were often a "normal" 6–7 weeks. The rationale was not the same as that for accelerated fractionation, which is intended to defeat rapidly proliferating tumours. This is described in the next section.

Dose per Fraction

There is now an overwhelming amount of radiobiological (Withers et al. 1982; Thames et al. 1982; Fowler 1984a) and clinical evidence (Singh 1978; Turesson and Notter, 1984a,b; Cox 1985) that if dose per fraction is increased the incidence of late damage rises, provided total doses are adjusted to give equal early damage. It is convenient to describe this difference in dose-per-fraction dependence of late and early damage in terms of the shape of the dose–response curve or even the underlying cell survival curve. The curves for late reacting tissues are more curvy than those for early reacting tissues, in the range of radiotherapy doses per fraction (Fowler 1983). The practical consequence of this leads to the modality of hyperfractionation by the same reasoning. If significantly smaller doses per fraction than usual are used (e.g. 1–1.5 Gy) there will be less late damage than usual when the total dose is adjusted to give equal early reactions (or equal tumour control).

Tumours have been shown to behave, in rats and mice, like rapidly proliferating tissues (Williams et al. 1985). This is the basis of hyperfractionation. Further, it is an obvious possibility to escalate the total dose in this modality until late reactions rise to the same incidence as usual, with a higher probability of tumour control and of early reactions. Which one of these two strategies is used with hyperfractionation depends on the relative magnitudes of early and late effects in normal tissues that are tolerable in the site and volume to be treated. To avoid increasing the overall time, more than one fraction per day must be used with hyperfractionation.

Thus both hyperfractionation and accelerated fractionation require more than one fraction per day, for entirely different reasons (Thames et al. 1983).

Factors Affecting Early Reactions

Early reactions occur from cell depletion in rapidly proliferating tissues. The time scale over which these reactions develop depends on the lifetime of the differentiated cells and their natural rate of replacement. Early reactions are seen within days in intestinal mucosa and within 3–4 weeks in mucosa and skin. Radiotherapy schedules can be prolonged or interrupted to avoid excessive damage to these tissues but tumour proliferation may then occur.

Early reactions can be reduced by:

1. Prolonging treatment to encourage proliferation to replace damaged cells.
2. Reducing the total dose (but see below).

Early reactions are increased by:

1. Shortening the overall time, and thus preventing compensatory proliferation.
2. Increasing the total dose. The dependence on total dose is not steep, however, if dose is incremented in daily fractions of about 2 Gy, because proliferation can almost keep up with this rate of killing cells in mucosa or skin.
3. Shortening the interval between fractions to the extent (i.e. less than 5–6 hours) that complete repair of sublethal damage cannot occur.

Tumours behave like early reacting, fast proliferating tissues in their response to fractionated radiotherapy. They may be spared, and then regrow, if the total dose is too small or the treatment time too long. They will be more effectively eliminated by high doses given over a short period of time, down to about a week as in interstitial implants. Reoxygenation can often be complete within this period. Acute reactions also increase with total dose and short overall times. Such reactions may therefore be a guide to tumour cell kill; they are *not* a guide to the severity of late reactions.

Factors Affecting Late Reactions

Late reactions occur in slowly proliferating tissues. These tissues are generally deeper than epithelial and the reactions may be life-threatening. Since they develop many months or years after the end of treatment, the schedule cannot be tailored or curtailed to avoid them in hypersensitive patients.

Late reactions are more severe if:

1. Doses larger than 2 Gy are used in each fraction. This results from the marked curvature of the dose–response curves for late responding tissues.
2. The total dose is increased, whether the overall time is lengthened or not.
3. The interval between fractions does not allow full repair of sublethal injury. This may require 6–8 hours for late reacting tissues (Fowler 1986b).

Late reactions are diminished if:

1. The total dose is kept low.
2. The dose per fraction is reduced below 2 Gy. This will result in more fractions being required. There is no disadvantage as regards late reactions if fractions are given twice daily, provided that the intervals between fractions are adequate.
3. The dose in every fraction is reduced somewhat (by partial shielding) rather than by applying a total shield after most of the fractions have been given. This is a constructive use of the steep dose-per-fraction response of late reacting tissues.

Adjustment of Total Dose in Non-standard Fractionation

Several isoeffect formulae exist. NSD, TDF and CRE are reasonable approximations for early reactions only, if the difference between the fractionation schedule being considered and a standard schedule is not too great (Orton and Ellis 1973). They are misleading for late effects (Fowler 1984c). If dose per fraction is increased the total dose must be reduced more dramatically than NSD, TDF or CRE would predict (Singh 1978; Cox 1985).

If the number (and therefore size) of fractions is to be changed without altering the overall time, there is a sounder biological basis in using the linear quadratic formula instead of NSD, TDF or CRE. The new dose can be calculated either for limiting early reacting tissues or for limiting late reacting tissues in the beams (see Appendix). Both calculations should normally be made so that a decision can be taken clinically about whether either reaction can be allowed to be any more severe than in conventional radiotherapy.

If the overall time is being altered two corrections may be necessary: one for complete or incomplete repair of sublethal damage between fractions, and another for changes in the amount of cell proliferation occurring within the treatment period. If intervals between fractions are at least 6 hours the first factor can be discounted; most tissues have a half-time of repair of 1–1.5 hours. The time course of repopulation is more complex, with a delay time followed by rapid proliferation as described above.

Shortening treatment times below 6 weeks will prevent the 4–6-week tissue recovery due to proliferation in skin, and the 3–6-week proliferation in mucosa. Early reactions to accelerated schedules should be more severe for this reason, but not the late reactions. An exception to this distinction occurs if the early reactions become so severe or long-lasting that they produce, as a consequence, a long-term tissue deficit. In principle very large decreases in total dose should be necessary on shortening from say 6 to 3 weeks overall time. In practice clinicians usually reduce the total dose by about 9 Gy to avoid the early reactions becoming too severe (range 3–10 Gy, median 8.7 Gy in the British Institute of Radiology's long versus short fractionation trial).

Appendix. The Linear Quadratic Formula for Calculating Isoeffective Doses

Changes in overall time must be allowed for separately

$$\text{Effect} = n\,(\alpha d + \beta d^2)$$

where n fractions of size d grays are given. The values α and β are the coefficients for the linear and the quadratic terms respectively. The ratio α/β (in grays) is equal to the dose at which the linear effect and the quadratic effect are exactly equal. If α/β is large the dose–response curve is nearly straight, so that over the radiotherapy range there is little change of total dose with dose per fraction. If α/β is small (2–4 Gy) the dose–response curve is rapidly curving in the region of radiotherapy dose per fraction. For early reacting tissues (and tumours) α/β = 10 Gy is often assumed, but this value is not critical. For late reacting tissues α/β is much more critical and values between 2 and 5 Gy have been found; α/β = 3 Gy is often assumed, but other values should be tested too.

In order to maintain equivalent biological effects (assuming no change in overall time) when going from n_1 fractions of size d_1 to n_2 fractions of size d_2, the following formula for prescribed dose should be used:

$$\text{Isoeffect} = n_1\,(\alpha d_1 + \beta d_1{}^2) = n_2\,(\alpha d_2 + \beta d_2{}^2)$$

from which, by simple algebra, we obtain:

$$\frac{\text{New total dose}}{\text{Old total dose}} = \frac{n_2 d_2}{n_1 d_1} = \frac{(\alpha/\beta + d_1)}{(\alpha/\beta + d_2} = \frac{1 + d_1/(\alpha/\beta)}{1 + d_2/(\alpha/\beta)}$$

These formulae are useful because estimates of the ratio α/β are available for many tissues although α or β separately are seldom known (Fowler 1984a,b; Hendry and Thames 1987).

The effect of overall time has to be considered separately, based on the considerations in the section 'Overall Time' above.

References

Cox JD (1985) Large-dose fractionation (hypofractionation). Cancer [Suppl] 55:2105–2111

Fowler JF (1983) Fractionation and therapeutic gain. In: Steel GG, Adams GE, Peckham MJ (eds) The biological basis of radiotherapy. Elsevier, Amsterdam, pp 181–194

Fowler JF (1984a) Review: total doses in fractionated radiotherapy: implications of new radiobiological data. Int J Radiat Biol 46:103–120

Fowler JF (1984b) Fractionated radiotherapy after Strandqvist. Acta Radiol [Oncol] 23:209–216

Fowler JF (1984c) Non-standard fractionation in radiotherapy (guest editorial). Int J Radiat Oncol Biol Phys 10:755–759

Fowler JF (1986a) Potential for increasing the differential response between tumours and normal tissues: can proliferation rate be used? Int J Radiat Oncol Biol Phys 12:641–645

Fowler JF (1986b) "Fading times" required for apparently complete repair in irradiated tissues assuming the linear quadratic model of dose response. Int J Radiat Biol 50:601–607

Hendry JH, Thames HD (1987) Fractionation in radiotherapy. Churchill Livingstone, Edinburgh

Orton CG, Ellis F (1973) A simplification in the use of the NSD concept in practical radiotherapy. Br J Radiol 46:529–537

Peters LJ, Ang KK (1986) In: Withers HR, Peters LJ (eds) Innovations in radiation oncology. Springer, Berlin Heidelberg New York

Saunders MI, Dische S (1986) Radiotherapy employing three fractions in each day over a continuous period of 12 days. Br J Radiol 59:523–525

Singh K (1978) Two regimes with the same TDF but differing morbidity used in the treatment of Stage III carcinoma of the cervix. Br J Radiol 51:357–362

Svoboda VHJ (1978) Further experience with radiotherapy by multiple daily sessions. Br J Radiol 51: 363–369

Thames HD, Withers HR, Peters LJ, Fletcher GH (1982) Changes in early and late radiation responses with altered dose fractionation: implications for dose–survival relationships. Int J Radiat Oncol Biol Phys 8:219–226

Thames HD, Peter LJ, Wither HR, Fletcher GH (1983) Accelerated fractionation vs hyperfractionation: rationales for several treatments per day. Int J Radiat Oncol Biol Phys 9:127–138

Trott KR, Kummermehr J (1985) What is known about tumour proliferation rates to choose between accelerated fractionation or hyperfractionation? Radiother Oncol 3:1–9

Turesson I, Notter G (1984a) The influence of fraction size in radiotherapy on the late normal tissue reaction. 1. Comparison of the effects of daily and once-a-week fractionation on human skin. Int J Radiat Oncol Biol Phys 10:593–594

Turesson I, Notter G (1984b) The influence of fraction size in radiotherapy on the late normal tissue reaction. 2. Comparison of the effects of daily and twice-a-week fractionation on human skin. Int J Radiat Oncol Biol Phys 10:599–606

van der Schueren E, Ang KK, Horiot JC, Gonzalez DG, Glabbeke M van, de Pauw M (1985) Concentrated radiotherapy schedules: role of repair and repopulation. In: Plenary session proceedings, 16th international congress of radiology, Hawaii, USA, 8–12 July, pp 99–104

Wang CC (1985) Accelerated fractionation. In: Plenary session proceedings, 16th international congress of radiology, Hawaii, USA, 8–12 July, pp 105–108

Williams MV, Denekamp J, Fowler JF (1985) A review of α/β ratios for experimental tumours: implications for clinical studies of altered fractionation. Int J Radiat Oncol Biol Phys 11:87–96

Wilson GD, McNally NJ, Dunphy E, Kärcher H, Pfranger R (1985) The labelling index of human and mouse tumours assessed by bromodeoxyuridine staining in vivo and in vitro and by flow cytometry. Cytometry 6:641–647

Withers HR, Thames HD, Peters LJ (1982) Differences in the fractionation response of acutely and late-responding tissues. In: Kaercher KH, Kogelnick HD, Reinartz G (eds) Progress in radio-oncology, vol 2. Raven Press, New York, pp 287–296

7 Fractionation in the Clinic

M. V. Williams

Introduction

Conventional radical radiotherapy employs 15–35 daily fractions delivered over 3–7 weeks. The results of regimens within this restricted range are broadly similar, and no consistent differences in tumour control or normal tissue damage have been demonstrated. Schedules with fewer fractions have been investigated because of the convenience they offer the patient. However, there is evidence that the therapeutic ratio is reduced if 10 or fewer fractions are used. Poor results have also been reported in several trials of split course treatment.

Recent research interest has focused on the use of a large number of smaller fractions often administered several times per day. This work has been encouraged by laboratory evidence that late responding normal tissues are more sensitive to dose per fraction than are tumours and early reacting normal tissues. Hyperfractionation aims to exploit this difference by giving more and smaller dose fractions.

Multiple fractions per day can also be used to reduce overall treatment time. It has been suggested that such accelerated fractionation might improve results by reducing tumour repopulation during treatment. Unfortunately, reduced repopulation in skin and mucosa during treatment leads to increased acute reactions. This necessitates a reduction in total dose which may offset any possible gain.

Dose–Time Relationships

The total dose of X-rays which can be tolerated by normal tissues varies with the organ irradiated and with treated volume, overall time, and the number and size of dose fractions. Early work on this problem concentrated on skin tolerance, which was then dose-limiting. The results were formalized by Ellis (1969) in his nominal standard dose (NSD) concept. This sought to provide a global variable for connective tissue tolerance, with only bone and brain being specifically excluded in the original paper.

The NSD was proposed as a clinical hypothesis and helped to stimulate laboratory research on dose fractionation. Normal tissue damage has been studied extensively in rodent and human tissues. Acutely responding normal tissues are those such as skin and mucosa which express the radiation injury shortly after it has been inflicted. They have a relatively short intermitotic cell cycle time and can therefore respond with a proliferative response to injury *during* a fractionated course of irradiation. Late responding normal tissues such as lung, kidney and spinal cord have a slow rate of cell turnover. Consequently radiation injury is expressed after a delay of several months. In addition these tissues are more sensitive to alterations in dose per fraction than are either acutely responding normal tissues or tumours (Fowler 1984). These results have led to suggestions that lower doses per fraction would improve the therapeutic ratio (Withers et al. 1982).

The Ellis formula provides a useful summary of isoeffective doses as regards acute skin tolerance for treatment delivered in 10–30 fractions over 4–6 weeks. Tumour response was not considered in detail, and the key question of how therapeutic ratio changes with dose fractionation was not addressed.

The problem of therapeutic ratio has been studied in experiments with transplanted rodent tumours: tumour control is a complex function of fraction number and size. The results can be understood in terms of the competing effects of reoxygenation rendering the remaining tumour cells more radiosensitive, and the repopulation occurring as overall time is prolonged making cure more difficult to achieve (Suit et al. 1977). Matching such tumour results to those obtained using normal tissue end-points allows an optimum regimen to be selected in these model systems. One of the main conclusions was that efficient reoxygenation is critical to success. Overall treatment time was also important because prolonged treatments were ineffective. The rate of clonogen production (potential doubling time) is similar in human and rodent tumours, although high cell loss results in lower macroscopic growth rates in man. This finding has led to clinical interest in reducing overall treatment time (for review see Fowler 1986 and Chap. 6 in this volume).

Biological Factors Underlying Dose Fractionation

Table 7.1 summarizes the clinical and biological factors to be considered when undertaking radical irradiation. The major physical factors over which the radiotherapist has control are radiation dose, fractionation and distribution. Careful planning can help to avoid damage to critical tissues such as the kidney, lung and spinal cord. Severe high-dose effects still occur and it has been suggested that these represent the extreme end of varying intrinsic radiosensitivity within the general population.

Table 7.1. Factors to be considered in radical irradiation

Host factors: the patient
Clinical
Age
Intercurrent illness
Size of tumour volume
Critical organs included
Biological
Repair capacity
Shape of dose–response curve
Repopulation
Redistribution
Tumour factors: the target
Clinical
Site
Histology
Size and stage
Macroscopic appearance
Regional metastases
Biological
Intrinsic radiosensitivity
Shape of dose–response curve
Hypoxic fraction
Reoxygenation pattern
Repopulation
Cell cycle redistribution

Clinically favourable indicators of tumour response such as an exophytic growth pattern are well established, but the underlying biological factors have not yet been investigated in the clinic. Tumours may recur locally for a variety of reasons. If these could be identified then treatment could be adjusted to the individual. For example accelerated fractionation could be given to rapidly proliferating tumours, and those which reoxygenate poorly could be treated with radiosensitizers. Such a selective approach would refine tumour groups and steepen dose–response curves, by grouping like with like. It should then be possible to define the role of novel approaches to radiotherapy.

Alternatives to Conventional Fractionation

Table 7.2 illustrates some of the different fractionation schemes which have been tested clinically and which will be discussed here. Conventional fractionation is illustrated as a course of 32 "daily" fractions delivering a total dose of 64 Gy over 44 days, such as might be used in the radical treatment of cancer of the bladder or of the head and neck. Field sizes and adjustments during treatment are not considered here, but are obviously critical.

Rapid fractionation delivering a radical dose of the order of 52 Gy in 16 fractions over 22 days has been practised for many years at the Christie Hospital,

Table 7.2. Some clinically tested radiation fractionation schemes

Fractionation regimen	Dose/ fraction (Gy)	No. of fractions	Total dose (Gy)	Fractionation scheme (weeks) 0 1 2 3 4 5 6 7 8
Conventional	2	32	64	XXXXX..XXXXX..XXXXX..XXXXX..XXXXX..XXXXX..XX
Rapid	3	16	52	XXXXX..XXXXX..XXXXX..X
Hypo-	4	10	45	X.X.X..X.X.X..X.X.X..X
Hyper-	2	63	76	XXXXX..XXXXX..XXXXX..XXXXX..XXXXX..XXXXX..XX XXXXX..XXXXX..XXXXX..XXXXX..XXXXX..XXXXX..X
Accelerated	2	30	60	XXXXX..XXXXX..XXXXX XXXXX..XXXXX..XXXXX
Fitz-	1.4	36	55	XXXXXXXXXXXX XXXXXXXXXXXX XXXXXXXXXXXX
Split course	2	30	60	XXXXX..XXXXX..XXXXX................XXXXX..XXXXX..XXXXX
Hyper- with split	1.3	84	84	XXXXX..XXXXX..XXXX.................XXXXX..XXXXX..XXXX XXXXX..XXXXX..XXXX.................XXXXX..XXXXX..XXXX XXXXX..XXXXX..XXXX.................XXXXX..XXXXX..XXXX

Manchester (Easson and Pointon 1985). Their treatment philosophy is unusual in that the irradiated volume is small and well defined, including only the primary tumour and its known extensions; prophylactic nodal irradiation is not usually undertaken. Case selection is rigidly defined and treatment planning and execution are extremely accurate. Late effects and tumour control appear comparable to those published elsewhere. Acute reactions occur earlier and are probably more severe, but are certainly of shorter duration. Such treatment can be considered as one extreme of "conventional"; at the other extreme are regimens which deliver 40 fractions of 1.8 Gy daily over 8 weeks. Acute reactions are mild but prolonged; there is no evidence that long-term results are any better.

Several alternative approaches to dose fractionation have been investigated in the clinic. Hypofractionated therapy administers fewer treatments, whereas hyperfractionation involves more and smaller fractions. Accelerated therapy uses multiple fractions per day in order to reduce the overall time. Fitzfractionation (accelerated hyperfractionation) uses both these concepts. Split course therapy introduces a rest period to allow recovery of acute normal tissue reactions. This approach has been combined with hyperfractionation.

Table 7.3 shows the five major alternatives to conventional radiation dose fractionation. All five approaches have a clear biological rationale in terms of tumour response. It is also evident that all could adversely effect the therapeutic ratio by causing increased normal tissue damage.

Criteria of an Adequate Clinical Trial

Clinical trials of fractionation can only be effective if they address a well-defined question. To ask if one treatment is better than another is imprecise. We need to know whether or not local tumour control has been improved and if more patients have been cured. It is important to show that the results of the new treatment are better than expected, rather than merely better than a poor control arm. Similarly we need to know whether or not the therapeutic ratio is improved, that is whether local tumour control is greater at a chosen level of late normal tissue damage. If not then a simple dose escalation using conventional fractionation should achieve the same result.

A study protocol must therefore include a control arm which is known to achieve good results and which is widely accepted as conventional. Survival, tumour control, and early and late normal tissue damage must be quantitated. Two dose levels are also needed in the novel arm of the trial to ensure that the result is interpretable (Fowler 1979). Fig. 7.1 illustrates the problem. The upper panel shows that the novel regimen results in higher tumour control rates, but normal tissue damage is also greater. The number of patients with uncomplicated local control increases from 30% (40%–10%) to 60% (80%–20%). This could simply be the result of an effective increase in dose, and therapeutic gain cannot be proved. The lower panel shows that the addition of a second dose group permits the conclusion that, at an equivalent level of normal tissue injury (10%), tumour control has been improved from 40% to 75%. In this example the lower-dose arm alone would have established therapeutic gain. A two dose

Table 7.3. Biological factors in unconventional fractionation

Fractionation regimen	Alteration	Rationale	Comment	Late effects	Early effects	Therapeutic gain?
Hypo-	Few, large fractions	Greater repair in tumour	Possibly melanoma and liposarcoma	Likely to be increased	Occur sooner and heal rapidly	None shown
Hyper-	Many, small fractions	Reduced repair in tumour	Supported by animal and human data	Must allow >4 hours for repair	Increased (less repair)	Possibly in bladder and head and neck tumours
Accelerated	Decreased overall time	Decreased tumour repopulation	Depends on potential doubling time	Must allow >4 hours for repair	Increased (less repopulation)	None in glioma
Fitz-	Smaller fractions in shorter time	Less repair and repopulation in tumour	Mixed rationale	Must allow >4 hours for repair	Increased (less repopulation and repair)	Not yet assessed
Split course	Split in therapy	Allows time for normal tissue recovery	Tumour can also repopulate	Increased if dose excessively increased	Reduced by rest period	None shown. Reduced in some studies

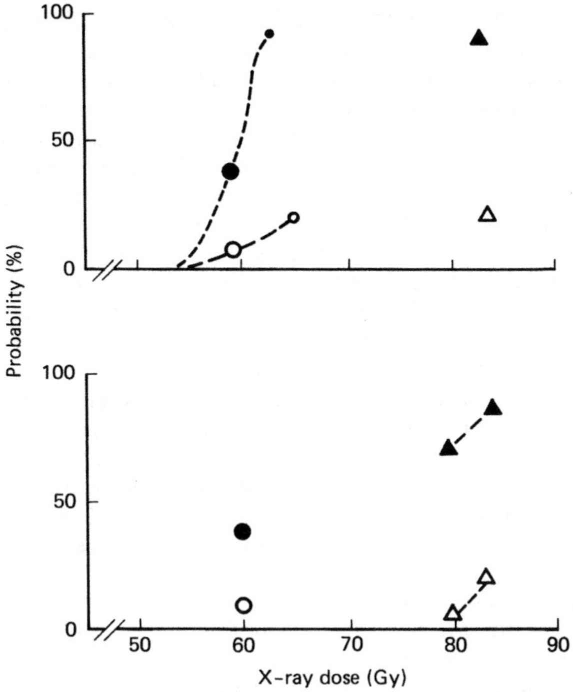

Fig. 7.1. Schematic results of a clinical trial of (for example) conventional irradiation (○) and hyperfractionation (△). Local control (*filled symbols*) and severe complications (*open symbols*) are shown.

In the *upper panel* a two-arm trial is shown. The novel treatment increases both local control and complications. One can argue that a simple increase in X-ray dose (*dashed line*) would have achieved the same result.

In the *lower panel* the use of two dose levels in the novel arm establishes that there is therapeutic gain. For a complication rate of 10%, local control increases from 40% to 75%.

level design thus doubles the chance of demonstrating such benefit. In practice the low incidence of rare events such as severe late normal tissue damage makes the confidence limits wide. The development of quantitative scoring systems for normal tissue injury should improve dose-sensitivity and reduce this problem.

Hypofractionation

Hypofractionation has been reviewed by Cox (1985), who concluded that it reduced the therapeutic ratio. He reported that tumour control was decreased and normal tissue damage increased. Hypofractionation should therefore be used with caution even when combined with new modalities such as radiosensitizers or hyperthermia.

Several important clinical trials of reduced fractionation in the radical treatment of head and neck tumours have been completed. Henk and James

(1978) treated 98 matched pairs of patients with either 10 fractions over 22 days or 30 fractions over 42 days. The patients were considered unfit for a trial involving hyperbaric oxygen and all had advanced tumours. The 3-year survival was 25% with no difference in tumour control or late effects in the two arms. Similar findings were reported by Weissberg et al. (1982) who treated a total of 64 patients with either 65 Gy in 32 daily fractions or 44 Gy in 11 fractions over 22 days. There was no difference in the results, but only 20% of the patients survived 3 years. Both studies involved relatively small numbers of patients so that a difference in results of 15% could have been missed. This criticism was particularly valid for normal tissue damage because of the small numbers of patients surviving to be at risk of late effects. One can conclude that for palliative treatment a few large fractions are adequate. These studies did not address the radical treatment of curable disease in fit patients.

The British Institute of Radiology undertook a study of 738 patients with advanced head and neck cancer treated either with 3 or 5 fractions per week (Wiernick et al. 1982). No significant differences between treatments were reported, but a wide range of fraction numbers and sizes were permitted in answering the pragmatic clinical question "Are three fractions a week better than five?" This aspect of trial design may have obscured important differences between fractionation schedules.

There have been no randomized trails of the effect of fraction size in treating carcinoma of the cervix, but in the Medical Research Council's studies of hyperbaric oxygen poor results were obtained by those centres using 6 or 10 fractions when compared with those using 20 or 27 fractions (Watson et al. 1978).

Split Course Irradiation

The initial rationale for a break during a course of radiotherapy was that it enabled a radical dose to be delivered to patients who were otherwise too frail to tolerate such treatment. It was believed that normal tissue recovery would occur faster than tumour regrowth, which is usually slow at the macroscopic level. It was also thought that a pause in treatment would allow tumour sensitization to the second course of irradiation by permitting redistribution of cells into sensitive phases of the cell cycle (Scanlon 1959).

Split course irradiation has been extensively investigated. There is no good evidence of therapeutic gain. Indeed tumour control may be decreased in head and neck tumours (Parsons et al. 1980a), and late damage increased in pelvic treatments, probably because improved acute tolerance allowed inappropriate dose escalation (Parsons et al. 1980b).

It is now thought that tumour proliferation is an important factor in determining local control. A break in treatment is therefore likely to be dangerous. Acute effects will be reduced but late normal tissue damage will not be lessened, because compensatory proliferation cannot occur in these tissues before damage has been expressed. These risks also apply to a break introduced to improve acute tolerance in a scheme involving multiple fractions per day. Tumour control may decrease and late normal tissue injury be unacceptable.

Accelerated Fractionation

Accelerated fractionation aims to decrease the time available for tumour repopulation. This has been shown to be an important factor in rodent tumours and, although human tumours grow more slowly, their potential doubling time can be as short as 3–10 days. For example 90% of squamous cell carcinomas have a doubling time of less than 7 days (Trott and Kummermehr 1985). If cell loss decreases during treatment then a 2–3-week reduction in overall treatment time might avoid three or four doublings in cell number, which would require a total fractionated dose of about 10 Gy in compensation (Thames et al. 1983). As with all novel treatments, translation of theory into increased local control and cure rates depends on the naturally occurring differences between tumours. If these are large then the dose–response curves will be shallow and benefit will be slight.

There have been many small pilot studies of accelerated fractionation in head and neck cancer. Tolerance diminishes as larger field sizes are used and is limited if the oral cavity is treated. In general a dose reduction of about 10 Gy is required if 30 fractions are given thrice daily in 14 days rather than daily over 42 days. Unfortunately, this approximates to the predicted reduction in tumour clonogens gained by reducing overall time. Such dose reductions, or alternatively the introduction of a break in treatment, may serve to neutralize the advantage which is being sought by using an accelerated regimen (for review see Thames et al. 1983).

Accelerated fractionation in the treatment of grade IV astrocytoma is of particular interest because acute reactions are not dose-limiting. Simpson and Platts (1976) delivered 40 Gy in 21 fractions over either 3 weeks or 7 days. There was no difference in survival, nor was there evidence of any increase in brain necrosis.

Hyperfractionation

The aim of a hyperfractionated regimen is to reduce the dose per fraction but not the overall time. The main rationale relates to proposed differences in the repair of radiation injury. The shape of a dose–effect curve can be described as the ratio of the α and β exponents in the equation:

$$\text{Surviving fraction} = e^{-(\alpha d+\beta d^2)}$$

In rodents, late responding normal tissues have low values of α/β (2–5 Gy), which implies a curvy dose–effect curve and a radiation response which will be greatly influenced by the size of the dose per fraction. Rodent tumours and human tumour cell lines have higher values of α/β (13–30 Gy). It should be possible to exploit this difference if the dose per fraction is reduced and total dose increased, so that late normal tissue injury is unchanged but tumour cell kill increased. Increased acute normal tissue injury is also to be expected (α/β of 7–15 Gy), but normal healing should occur.

It has also been suggested that at low doses per fraction the oxygen effect may be less important. Differences in rates of repair may also confer therapeutic gain (Douglas 1982). There is little in vivo evidence to support these speculations.

If hyperfractionation is to be advantageous, total dose must be increased to take advantage of differences in repair (Thames et al. 1983). In a study on the treatment of bladder cancer Edsmyr et al. (1985) increased the total dose to 84 Gy given in twice-daily fractions of 1 Gy. The control arm received 64 Gy in 32 daily fractions; and in both regimens a 2-week break in treatment was used to spare acute effects. In an analysis combining T_2 and T_3 tumours local control and survival were improved. However, the finding of only 34% of patients alive at 5 years was no better than the figure of 30% frequently reported with conventional irradiation without a split course. One can conclude that hyperfractionation improved on a poor regime and that the split course regimen probably allowed tumour proliferation.

Retrospective studies on the treatment of head and neck cancer have been reported by a number of authors. Parsons et al. (1984) found no improvement in results but considered that the shorter hospital stay might be advantageous for both patient and radiotherapist. Knee et al. (1985) reported on the use of concurrent boost therapy and considered that the results were better than might have been expected. Similarly Wang et al. (1986) claimed improved results compared with their previous experience. Formal clinical trials are required to establish whether or not there is therapeutic gain.

Fitzfractionation

Fitzfractionation is the term suggested by Withers et al. (1982) for a regimen which has no single biological rationale but which seeks to combine the virtues of hyperfractionation and acceleration by using multiple daily fractions of less than 2 Gy in a shorter overall time. If this proves to be an effective way to treat patients it will not matter that we do not know why it is better.

Nguyen et al. (1985) reported a study of accelerated split course hyperfractionation in the treatment of advanced head and neck cancer. Eight fractions per day of 0.9 Gy were administered for a week and the treatment repeated after a 2-week rest. A total dose of 72 Gy was delivered in 80 fractions over 26 days. Severe acute normal tissue damage with necrosis and haemorrhage occurred in 34% of the patients. Tumour control was poor and severe late complications were seen in 80% of the survivors. These unacceptable results probably reflect incomplete repair of radiation damage in the 2-hour interval between the fractions. Thames et al. (1984) have shown that the half-time for repair differs between acutely responding tissues (0.3–0.9 hours) and late responding tissues (1.5 hours). It is likely that this will prove a limiting factor in the clinical application of hyperfractionation, as calculations by Dale (1986) have shown.

The European Organization for Research and Treatment of Cancer (EORTC) has conducted a large randomized trial of accelerated split course hyperfractionation in head and neck cancer. Three fractions a day of 1.6 Gy were delivered. The first part of the course consisted of 30 fractions over 12 elapsed days; after a 3–4-week rest a further 15 fractions were delivered over

5 days. A total dose of 72 Gy was delivered in 45 fractions over about 40 days. This was compared with 70 Gy in 35 daily fractions of 2 Gy delivered over 47 days. Preliminary results show no difference in survival, locoregional control or late effects (van den Bogaert et al. 1986). This is perhaps not surprising if one considers that the total dose, number of fractions and overall treatment time are similar, despite the differences in fractionation.

Saunders et al. (1986) treated non-small-cell lung cancer using a very short regimen with small dose fractions: three fractions per day of 1.4 Gy were given, to a total dose of 50.4 Gy in 36 fractions over 12 days. Unusually, the treatment was continued over the weekend. They were careful to ensure a 6-hour interval between fractions. Acute reactions occurred sooner than with conventional irradiation, but were tolerable. Initial tumour control rates were high and long-term results are awaited with interest.

Shin et al. (1985) have reported a trial of accelerated hyperfractionation in the treatment of malignant cerebral astrocytoma. Patients in the control arm received 58 Gy in 30 daily fractions over 6 weeks. The novel treatment consisted of 61.4 Gy in 69 fractions of 0.9 Gy three times a day at intervals of 3–4 hours over 4.5 weeks. One-year survival was significantly improved from 20% to 41%, and the authors claimed that their control results were as good as those in other series. Since the only toxicity was a slight increase in skin reactions with no severe late effects, the next step is to increase the dose. Animal results imply that this should be possible, but it should be remembered that much of the data relates to spinal cord rather than brain, that there are few data for doses below 2 Gy, and that the data that are available showed that repair was less than that predicted by the linear quadratic model (Ang et al. 1985). If the repair half-time is indeed long in neural tissue then fraction intervals of at least 6 hours would be advisable (Dale 1986). Shin and his colleagues are continuing their work with a dose escalation to 71.2 Gy in 80 fractions of 0.9 Gy over 5.5 weeks.

Predictive Sensitivity Testing

In clinical practice little is known about any of the biological factors (Table 7.1) which determine tumour and normal tissue responses in an individual. It would be a major advance if we were able to study these prospectively and choose the optimum fractionation regimen for an individual.

Predictive radiosensitivity testing is now being attempted clinically (see review by Peters et al. 1986). The shape of dose–response curves will need to be reliably characterized in order to be effective in determining the optimum fractionation. Tumour oxygenation can be studied using oxygen electrodes or by frozen tissue biopsies in which the oxygenation status of red cell haemoglobin is determined. Nuclear magnetic resonance spectroscopy carries the promise of a non-invasive tool for such investigations. Tumour cell kinetics can now be studied in man by obtaining biopsies after the administration of BrdU, detecting its presence in cells using a monoclonal antibody. The potential tumour doubling time can then be determined and could be prospectively investigated to establish whether or not it predicts a favourable response to an accelerated course of treatment.

Normal tissue radiation response can also be investigated. In the future fibroblast culture and radiation sensitivity testing might permit the identification of patients at high risk of serious late effects. Dose escalation might then be possible for other patients, with a consequent improvement in tumour control rates.

If some or all of these aspects of tumour and normal tissue response could be studied it might be possible to use experimental modalities such as altered fractionation, radiosensitizers, radioprotectors and fast neutrons more rationally. At present such novel treatments are applied blindly to patients identified as having a poor prognosis, with no stratification according to the underlying reason for treatment failure. Peters et al. (1982) have pointed out that such a mixed population results in shallow dose–response curves and diminishes the chance of identifying a genuine advance in treatment. All modifications to radiotherapy depend on the assumption that dose–response curves are steep. If they are not then any innovation which depends on modifying the therapeutic ratio will have little impact.

Conclusions

Reduced fractionation and split course regimens can give poor results and have certainly never been shown to improve the therapeutic ratio even if they are more cost-effective.

The use of multiple daily fractions has excited much recent interest, but there is no conclusive evidence of benefit. Phase III clinical trials are now required to test the hypothesis that these novel approaches to fractionation can improve the therapeutic ratio. Two dose levels should be used in the novel arm to ensure an interpretable result.

Large long-term studies are required: further small studies will not help to answer the important questions. Parallel investigation of the biological properties of tumours and normal tissues in an attempt to individualize treatment on a rational basis are likely to make such studies more useful.

References

Ang KA, van der Kogel AJ, van der Schueren E (1985) Lack of evidence for increased tolerance of rat spinal cord with decreasing fraction doses below 2 Gy. Int J Radiat Oncol Biol Phys 11:105–110

Cox JD (1985) Large-dose fractionation. Cancer 55:2105–2111

Dale RG (1986) The application of the linear-quadratic model to fractionated radiotherapy when there is incomplete normal tissue recovery between fractions, and possible implications for treatments involving multiple fractions per day. Br J Radiol 59:919–927

Douglas BG (1982) Superfractionation: its rationale and anticipated benefits. Int J Radiat Oncol Biol Phys 8:1143–1153

Easson EC, Pointon RCS (eds) (1985) The radiotherapy of malignant disease. Springer, Berlin Heidelberg New York

Edsmyr F, Andersson I, Esposti PL, Littbrand B, Nilsson B (1985) Irradiation therapy with multiple small fractions per day in urinary bladder cancer. Radiother Oncol 4:197–203

Ellis F (1969) Dose, time and fractionation: a clinical hypothesis. Clin Radiol 20:1–7

Fowler JF (1979) Doses and fractionation schemes to be employed in clinical trials of high-LET radiations. In: Barendsen GW, Broerse J, Breur K (eds) High-LET radiations in clinical radiotherapy. Pergamon, Oxford, pp 263–266

Fowler JF (1984) Review: total doses in fractionated radiotherapy – implications of new radiobiological data. Int J Radiat Biol 46:103–120

Fowler JF (1986) Potential for increasing the differential response between tumors and normal tissues: can proliferation rate be used? Int J Radiat Oncol Biol Phys 12:641–645

Henk JM, James KW (1978) Comparative trial of large and small fractions in the radiotherapy of head and neck cancer. Clin Radiol 29:611–616

Knee R, Fields RS, Peters LJ (1985) Concomitant boost radiotherapy for advanced squamous cell carcinoma of the head and neck. Radiother Oncol 4:1–7

Nguyen TD, Demange L. Froissart D, Panis X, Loirette M (1985) Rapid hyperfractionation radiotherapy. Clinical results in 178 advanced squamous cell carcinomas of the head and neck. Cancer 56:16–19

Parsons JT, Bova FJ, Million RR (1980a) A re-evaluation of split-course technique for squamous cell carcinoma of the head and neck. Int J Radiat Oncol Biol Phys 6:1645–1652

Parsons JT, Thar TL, Bova FJ, Million RR (1980b) An evaluation of split-course irradiation for pelvic malignancies. Int J Radiat Oncol Biol Phys 6:175–181

Parsons JT, Cassisi NJ, Million RR (1984) Results of twice-a-day irradiation of squamous cell carcinomas of the head and neck. Int J Radiat Oncol Biol Phys 10:2041–2051

Peters LJ, Withers HR, Thames HD, Fletcher GH (1982) The problem: tumour radioresistance in clinical radiotherapy. Int J Radiat Oncol Biol Phys 8:101–108

Peters LJ, Brock WA, Johnson T, Meyn RE, Tofilon PJ, Milas L (1986) Potential methods for predicting tumor radiocurability. Int J Radiat Oncol Biol Phys 12:459–467

Saunders MI, Dische S (1986) Radiotherapy employing three fractions in each day over a continuous period of 12 days. Br J Radiol 59:523–525

Scanlon PW (1959) The effect of mitotic suppression and recovery after irradiation on time–dose relationships and the application of this effect to clinical radiation therapy. Am J Roentgenol 81:433–455

Shin KH, Urtasun RC, Fulton D et al. (1985) Multiple daily fractionated radiation therapy and misonidazole in the management of malignant astrocytoma. Cancer 56:758–760

Simpson WJ, Platts ME (1976) Fractionation study in the treatment of glioblastoma multiforme. Int J Radiat Oncol Biol Phys 1:639–644

Suit HD, Howes AE, Hunter N (1977) Dependence of response of a C3H mammary carcinoma to fractionated irradiation on fractionation number and intertreatment interval. Radiat Res 72:440–454

Thames HD, Peters LJ, Withers HR, Fletcher GH (1983) Accelerated fractionation vs hyperfractionation: rationales for several treatments per day. Int J Radiat Oncol Biol Phys 9:127–138

Thames HD, Withers HR, Peters LJ (1984) Tissue repair capacity and repair kinetics deduced from multifractionated or continuous irradiation regimens with incomplete repair. Br J Cancer 49 [Suppl VI]:263–269

Trott K-R, Kummermehr J (1985) What is known about tumour proliferation rates to choose between accelerated fractionation or hyperfractionation? Radiother Oncol 3:1–9

Van den Bogaert W, van der Schueren E, Horiot J-C et al. (1986) Early results of the EORTC randomized clinical trial on multiple fractions per day (MFD) and misonidazole in advanced head and neck cancer. Int J Radiat Oncol Biol Phys 12:587–591

Wang CC, Suit HD, Blitzer PH (1986) Twice a day radiation therapy for supraglottic carcinoma. Int J Radiat Oncol Biol Phys 12:3–7

Watson ER, Halnan KE, Dische S et al. (1978) Hyperbaric oxygen and radiotherapy: a Medical Research Council trial in carcinoma of the cervix. Br J Radiol 51:879–887

Weissberg JB, Son YH, Percarpio B, Fischer JJ (1982) Randomized trial of conventional versus high fractional dose radiation therapy in the treatment of advanced head and neck cancer. Int J Radiat Oncol Biol Phys 8:179–185

Wiernick G, Bates TD, Berry RJ et al. (1982) Seventh interim progress report of the British Institute of Radiology fractionation study of 3F/week versus 5F/week in radiotherapy of the laryngopharynx. Br J Radiol 55:505–510

Withers HR, Peters LJ, Thames HD, Fletcher GH (1982) Hyperfractionation. Int J Radiat Oncol Biol Phys 8:1807–1809

8 Radiation Resistance and the Dose-Rate Effect: Experimental

G. G. Steel and A. Horwich

Introduction

Prior to 1980 it was widely thought that radiation is a relatively non-specific killing agent, in the sense that it was not possible to attribute success or failure in clinical radiotherapy to differences in the inherent radiosensitivity of tumour cells. For some decades, therefore, radiobiological research has concentrated mainly on hypoxia: it was tacitly believed that failure in treatment is usually related to the presence of hypoxic cells, which are known to be considerably more resistant to low linear energy transfer (LET) radiation than well-oxygenated cells. Reoxygenation of hypoxic cells is known to occur in experimental tumours and one reason for clinical failure could be the failure of a tumour to reoxygenate satisfactorily during a conventional course of radiotherapy.

Through the original work of Fertil and Malaise (1981), and the confirmatory studies of Deacon et al. (1984), it has become clear that this view of clinical failure may not be accurate. These two studies have shown that among those human tumour cell lines that have so far been investigated there are considerable differences in the radiosensitivity of oxic cells. This is manifested not in the final (i.e. high-dose) slope of the cell survival curve, but in its initial or low-dose region. Furthermore there is evidence for a positive correlation between this initial slope and the clinical response to radiotherapy of various categories of tumour. The implication is that inherent cellular radiosensitivity is an important determinant of clinical outcome. This does not refute the idea that hypoxia limits the response of some tumours, but it points towards another quite different possible cause of treatment failure.

The initial slope of the oxic cell survival curve is probably due to the induction of unrepairable radiation damage. It is likely that radiation exposure induces a number of discrete lesions within the cell. On the original basis of "target theory" it was thought that each cell has a critical number of targets, all of which must be inactivated to kill the cell. More recently it has been supposed that irradiation gives rise to a mixture of lethal and non-lethal events; an additional radiation dose may convert non-lethal into lethal lesions. Immediately after irradiation, repair processes begin to operate and the persistence of only one unrepaired lethal event is by definition sufficient to kill the cell. Both of these models provide an explanation of the shoulder that is usually observed on cell survival curves, but the repair-type models lead to the view that the explanation of why some cells have a steeper initial slope than others is to be sought in their sensitivity to repairable and lethal damage.

This line of thought encourages more detailed study of the inherent radiosensitivity of human tumour cells. We need more accurate information on the initial part of the cell survival curve. Furthermore, to the extent that repair is implicated in clinical failure there is a need for extensive study of repair-modifying agents.

Low-Dose-Rate Irradiation

The study of low-dose-rate irradiation is of interest from both a fundamental and an applied clinical point of view. Low-dose-rate irradiation is the extreme case of increased fraction number and reduced dose per fraction. It allows maximal recovery from radiation damage, for a given duration of treatment, and the lower the dose rate the greater the opportunity for recovery. Its attraction in the present context is that an exploration of the dose-rate effect should give information on both the extent and the kinetics of recovery from radiation damage.

The clinical use of low-dose-rate radiotherapy has a long history. It has most commonly been used in interstitial or intracavitary treatment, but since these treatments have mainly been designed to exploit the geometric advantages of the very non-uniform fields around radioactive sources, it has been difficult to deduce whether the low dose rate by itself carries a therapeutic advantage (Hall 1978).

Clinical acute radiation treatments are usually given within a few minutes. If the treatment time is prolonged by reducing the dose rate a number of biological processes are permitted to take place during treatment, thus modifying the radiation effect. Four processes are of particular importance, which have been called the four Rs of radiobiology: repair, reassortment, reoxygenation and repopulation (Withers 1975).

The range of dose rates over which any of these processes will modify response depends on the speed of the process. Repair is the fastest, associated with a half-time of perhaps an hour or less; reassortment will occur within the approximate duration of the cell cycle phases (a few hours); repopulation requires one or more cell cycle times (a few days in human tumour cells); the speed of reoxygenation in human tumours is difficult to specify but it may well vary over the range described for reassortment and repopulation. Any particular process will modify response whenever the duration of treatment becomes comparable with the half-time of the process. Fast processes (short half-times) will be able to compete with a rapid infliction of damage (high dose rate). Conversely, a slow process will only influence response at low dose rate.

The irradiation time for a dose of 2 Gy is 2 min at 100 cGy/min, 20 min at 10 cGy/min, and 200 min (i.e. 3.3 h) at 1 cGy/min. As we have argued previously (Steel et al. 1986) the dose-rate effect down to about 2 cGy/min will therefore be dominated by repair processes. At around 2 cGy/min cell cycle progression phenomena may begin to modify response. Nevertheless, studies of the sparing effect of reducing the dose rate from say 100 to 2 cGy/min can give information on the kinetics of cellular recovery from radiation damage.

Mathematical Models of the Dose-Rate Effect

Most models of cell survival after radiation allow the calculation of response as a function of dose but do not take account of dose rate. It is quite difficult to adapt the multitarget or linear-quadratic equations in a rational way to deal with changes in dose rate. Oliver (1964) suggested an empirical way of doing this which envisaged that the dose-equivalent of recoverable damage should decay exponentially. This approach has more recently been used by Thames (1985) in his "incomplete repair" model.

The "lethal–potentially lethal" (LPL) model of Curtis (1986) is conceptually different. It is a kinetic model of cell killing by radiation in which two types of lesion are thought to occur: lethal (irreparable) lesions and potentially lethal lesions. The potentially lethal lesions may either be repaired to restore cell viability, or binary interaction may convert them into lethal lesions so that they are thus "fixed". It is envisaged that any one unrepaired lesion (of either type) will lead to cell death. We have compared these two models in recent publications (Steel et al. 1986, 1987; Stephens et al. 1987). In general they fit the data equally well and they similarly describe the dose-sparing that is seen over the range 200–2 cGy/min.

The Dose-Rate Effect in Human Tumour Cell Lines

Studies at the Institute of Cancer Research, London, have been made on over 17 human tumour cell lines of a variety of histologies, and the results so far have recently been reviewed (Steel et al. 1987). Most of the cell lines were derived from tumours that had first been xenografted into immune-deficient mice. Radiobiological experiments were performed on single cell suspensions prepared by enzymatic disaggregation of xenografts or confluent cell culture monolayers. In all cases, a standard protocol was followed before, during and after irradiation. Cells were plated out at appropriate density in soft agar culture or on plastic in monolayer, and held for 2 h at 37 °C in a 5% O_2/5% CO_2/90% N_2 gaseous environment before irradiation with ^{60}CO γ-rays at the appropriate

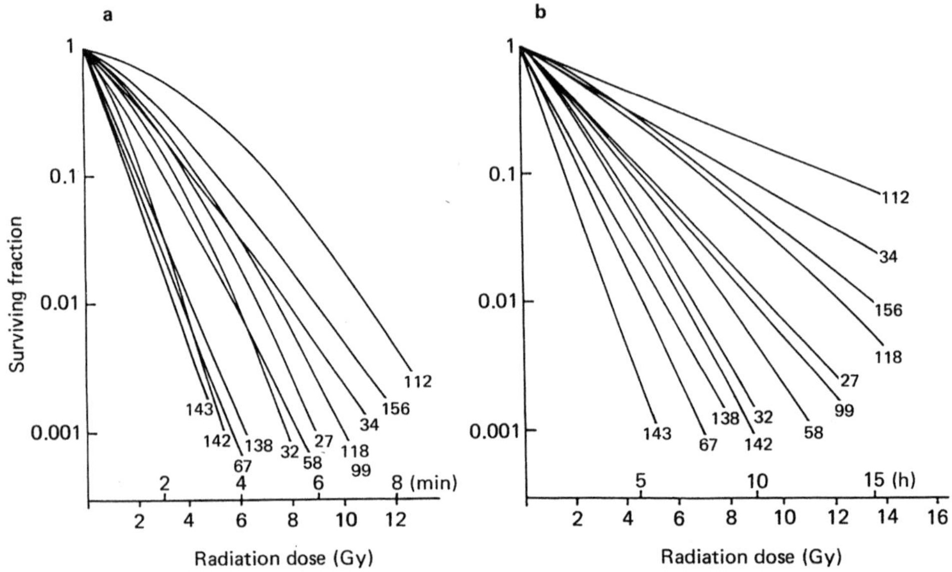

Fig. 8.1a,b. Cell survival curves for 12 human tumour cell lines irradiated at high dose rate (**a**: approx. 150 cGy/min) or at low dose rate (**b**: approx. 1.6 cGy/min). The duration of exposure has been marked on the abscissa. The curves are labelled with a number that denotes the cell line: 34 and 118, melanoma; 32 and 58, pancreas; 99, breast; 156, cervix; 67 and 112, bladder; 27, teratoma; 138, 142 and 143, neuroblastoma.

dose rate. Cells were maintained at 37 °C and in the low oxygen atmosphere throughout irradiation and during the post-irradiation period of colony growth.

The acute radiation survival curves of the 12 cell lines on which dose-rate data are available are summarized in Fig. 8.1a. Although there is a range of radiosensitivities it can be seen that the final slopes of these acute curves do not vary widely (in fact by a factor of 2.0). The major difference is in the steepness of the initial slope, which varies by a factor of about 12.

Fig. 8.1b shows the corresponding range of radiosensitivity at low dose rate (1.6 cGy/min). At this dose rate most, but not all, of the survival curves have become exponential. The curves that were steepest at high dose rate have not moved much, but as expected the survival curves for the more resistant lines have flattened considerably. The final slopes now differ by a factor of 7. It can be seen that low-dose-rate irradiation discriminates better than high-dose-rate radiation between tumour cell lines of differing radiosensitivity.

The sparing effect of low-dose-rate irradiation can usefully be described by the type of representation given in Fig. 8.2. For each cell line and at each dose rate the radiation dose required to give 1% cell survival has been determined (and called the $D_{0.01}$ value). This has been done by fitting each data set individually with a linear-quadratic equation and thus interpolating the $D_{0.01}$ value. The dose reduction factor (DRF) indicates the ratio of $D_{0.01}$ at a specified dose rate to $D_{0.01}$ at high dose rate. The term DRF is used by analogy with the terms dose enhancement factor or sensitizer enhancement ratio that are used when there is

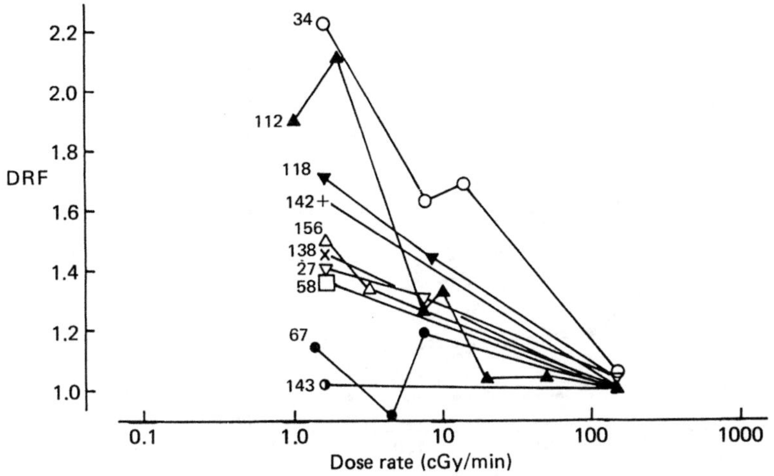

Fig. 8.2. Low-dose-rate sparing as a function of dose rate in human cell lines. DRF indicates the dose required at the specified dose rate to produce 10^{-2} survival divided by the dose to produce this effect at infinitely high dose rate. The curves are numbered as in the legend to Fig. 8.1.

a change towards steeper survival curves. There clearly is a wide range among these cell lines in the amount of sparing associated with a lowering of dose rate. Two cell lines (143 neuroblastoma, 67 bladder cancer) show no significant sparing; in other lines the dose-sparing at 2 cGy/min ranges up to a factor of 2. A similar range of sparing factors has been found for in vitro cell lines of non-human origin (Steel et al. 1986).

Dose-Rate Effects in Normal Tissues

A number of studies have investigated the change in response of normal tissues in the mouse as dose rate is reduced, and it is of interest to compare the results on the same type of plot used in Fig. 8.2 for the human tumour cell data. Fig. 8.3 is taken from a recent review of dose-rate effects (Steel et al. 1986), and includes two types of data. Studies where the end-point was gross tissue failure are indicated by continuous lines on the scale labelled ED_{50}, where ED_{50} is the radiation dose at any given dose rate that is required to produce the effect in 50% of animals. Some studies in the bone marrow and intestine have employed cell cloning techniques and the results of these (D_0 values) are indicated by broken lines in the figure. Also included are the results of a more recent cell survival study on the mouse testis (Delic et al. 1987).

The results reveal a wide range of sparing effects for low-dose-rate irradiation. The studies in bone marrow failed to detect any significant sparing. At the other extreme, the lung and hair follicle show considerable sparing; for these tissues the ED_{50} at 2 cGy/min is about 1.7 times the value at high dose rate.

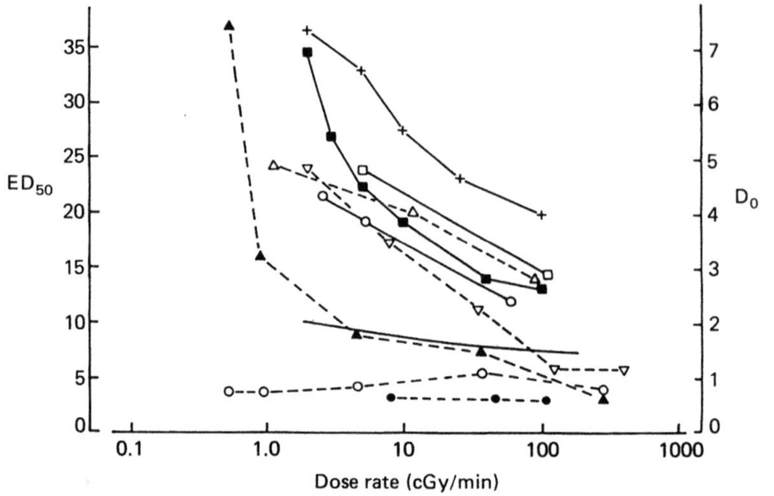

Fig. 8.3. Summary of dose-rate dependence of radiation damage to murine normal tissue. The *continuous lines* show data on ED_{50} in Gy (*left-hand scale*): ■——■, radiation pneumonitis; □——□, late lung damage; ○——○, intestinal damage; +——+, epilation. The *continuous line without points* is representative of nine published studies of 30-day mortality following whole body irradiation. The *broken lines* indicate D_0 values (Gy) obtained by in vivo cell cloning: ▽---▽ and ▲---▲, two studies on the intestine; ○---○ and ●---●, bone marrow CFU-S; △---△, the testis. (See Steel et al. (1986) for sources. The data on the testis are from Delic et al. (1987).)

Clinical Implications of the Low-Dose-Rate Studies

It is clear from the data presented in Fig. 8.2 that substantial sparing of radiation damage is observed in some human tumour cell lines as the dose rate is reduced to around 2 cGy/min. Fig. 8.3 shows a very similar picture among a small number of mouse normal tissues. The greatest degree of sparing among the normal tissues is closely matched by the sparing seen in some tumour lines. It is therefore not possible to claim that in general the use of low dose rate in clinical radiotherapy should lead to a therapeutic advantage. It would seem from the wide range of sparing seen in these human tumour cell lines that, depending upon the precise circumstances, a lowering of dose rate could lead either to a therapeutic advantage or to a therapeutic disadvantage. Tumour cell types that show little sparing, growing in an anatomical site where the dose-limiting normal tissues have a high degree of sparing, will be better treated by low-dose-rate therapy. This may well be the case, for instance, with total body irradiation in the preparation of leukaemic patients for bone marrow grafting; low-dose-rate irradiation should be effective in eradicating the leukaemic cells (and the marrow) but it should spare damage to most other normal tissues. However, the converse could also be the case. Radioresistant tumours that are well spared by low dose rate could be at an advantage if the limiting normal tissues have a poor capacity to recover. An encouraging thought is that it may well be that late reacting normal tissues have a generally high capacity to recover from radiation damage (Thames et al. 1982; Fowler 1984) and should therefore tend to show a relatively high degree of low-dose-rate sparing. More attention needs to be given to selection of those tumours that will respond best to low-dose-rate therapy, and of course this must be done with due consideration to the dose-limiting normal tissues.

The experimental studies reviewed here are the work of a number of our past and present colleagues: Judy Deacon, Julian Delic, Julian Down, Gill Duchesne, Lloyd Kelland and John Peacock. We are most grateful to them for allowing us to reproduce their data.

References

Curtis SB (1986) Lethal and potentially lethal lesions induced by radiation – a unified repair model. Radiat Res 106:252–270

Deacon JM, Peckham MJ, Steel GG (1984) The radioresponsiveness of human tumours and the initial slope of the cell survival curve. Radiother Oncol 2:317–323

Delic JI, Schlappack OK, Steel GG (1987) Effect of dose rate on the survival of murine spermatogonia following ^{60}Co γ-irradiation. Radiother Oncol 8:345–351

Fertil B, Malaise EP (1981) Inherent cellular radiosensitivity as a basic concept for human tumour radiotherapy. Int J Radiat Oncol Biol Phys 7:621–629

Fowler JF (1984) What next in fractionated radiotherapy? The First James Kirk Memorial Lecture. Br J Cancer 49 [Suppl VI]: 285–300

Hall EJ (1978) The promise of low dose rate: has it been realized? Int J Radiat Oncol Biol Phys 4:749–750

Oliver R (1964) A comparison of the effects of acute and protracted gamma-radiation on the growth of seedlings of *Vicia faba*. II. Theoretical calculations. Int J Radiat Biol 8:475–488

Steel GG, Down JD, Peacock JH, Stephens TC (1986) Dose-rate effects and the repair of radiation damage. Radiother Oncol 5:321–331

Steel GG, Deacon JM, Duchesne GM, Horwich A, Kelland LR, Peacock JH (1987) The dose-rate effect in human tumour cells. Radiother Oncol 9:299–310

Stephens TC, Eady JJ, Peacock JH, Steel GG (1987) Split-dose and low dose-rate recovery in four experimental tumour systems. Int J Radiat Biol 52:157–170

Thames HD (1985) An "incomplete-repair" model for survival after fractionated and continuous irradiations. Int J Radiat Biol 47:319–339

Thames HD, Withers HR, Peters LJ, Fletcher GH (1982) Changes in early and late radiation responses with altered dose fractionation: implications for dose–survival relationships. Int J Radiat Oncol Biol Phys 8:219–226

Withers HR (1975) The four Rs of radiotherapy. Adv Radiat Biol 5:241–271

9 Low Dose Rate in Clinical Radiotherapy

A. Horwich, P. Blake and G. G. Steel

Introduction

The radiobiological considerations underlying the cytotoxicity of low-dose-rate irradiation have been discussed in the preceding chapter. In summary they consist of repair during irradiation, which decreases cytotoxicity, and cell cycle progression which would be expected to increase cytotoxicity by virtue of resistant surviving cells moving into a more sensitive phase of the cell cycle (Table 9.1). Proliferation during irradiation may reduce the effects of low-dose-rate treatment. However, the important factor restricting repopulation is the overall treatment duration, and in many clinical examples this may be shorter after continuous low-dose-rate radiotherapy than after fractionated treatment at conventional dose rate.

Table 9.1. Radiobiology of low-dose-rate irradiation

Parameter affected	Low-dose-rate cytotoxicity
Sublethal damage repair	↓
Cell cycle progression	↑
Cell proliferation	↓
Reoxygenation	↓ ↑

Prescribed dose rate in clinical radiotherapy ranges from 0.1 cGy/min in iodine-125 implants up to approximately 400 cGy/min from linear accelerators. The major intermediate groupings include interstitial implants and intracavitary insertions at 0.5–2 cGy/min and total body irradiation at ⩾2 cGy/min.

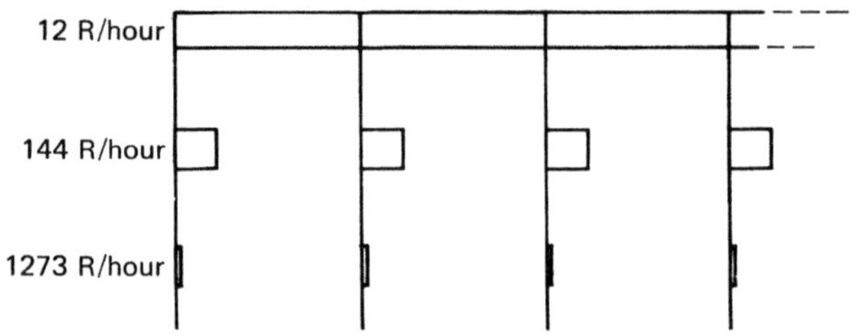

Fig. 9.1. Study design for comparison of dose-rate effects on normal skin (McWhirter 1936). See text for details.

The first proponent of low-dose-rate radiotherapy was Claudius Regaud (1870–1940). While working at the Radium Institute in Paris at a time when the conventional radiotherapy of cancers consisted of a single large dose, he proposed a smaller total dose protracted by continuous treatment for 6–15 days. He believed that this technique would permit the use of radiotherapy as a selectively toxic rather than a "cyto caustic" treatment modality (Regaud 1922). Clinical analysis of dose-rate effects was undertaken by McWhirter (1936), who observed the skin reactions in human volunteers irradiated by radium moulds symmetrically on each thigh at different dose rates. In a study treating the same area on each thigh of 16 patients to between 2000 and 2300 roentgens delivered in one fraction per day for 7 days (Fig. 9.1) he was unable to observe any difference between dose rates ranging from 12 R/hour to 1273 R/hour (the former representing continuous irradiation over the 7 days). In contrast, he demonstrated a lower dose of 2000 R required for skin tumour control during 2-hour treatment compared with 4400 R required for tumour control when using 7-day continuous treatment. These data are supported by those of Mitchell (1960) (Table 9.2).

Table 9.2. Doses required for control of small skin tumours treated at a range of dose rates

Dose rate (cGy/min)	Total dose required (CGy)	Treatment duration (hours)
18.0	1800	1.67
3.7	2250	10
2.0	2880	24
1.13	3300	48.5
0.85	3600	70.3
0.45	4500	169
0.23	4800	336
0.13	4950	600

From Mitchell (1960).

In the 1930s and 1940s interstitial implant techniques with radium were used successfully in a range of clinical situations, especially oropharyngeal tumours. Clinical observations supported a dose-rate effect and Paterson (1952) and Ellis (1968) published isoeffect curves suggesting change in the total interstitial implant dose from 6000 cGy for situations where the dose rate differed from the

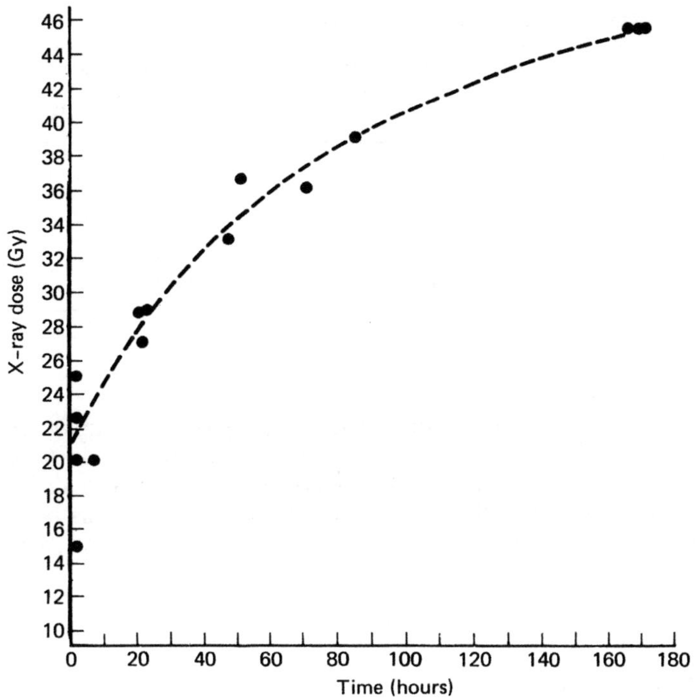

Fig. 9.2. Isoeffect curve for oral cavity implants at different dose rates (After Paterson 1952.)

standard (0.59 cGy/min) (Fig. 9.2). Subsequently, this form of isoeffect relationship has been modelled mathematically and used to derive corrections for different dose rates (Kirk et al. 1972; Orton and Ellis 1973).

Formulae describing dose-rate effects have also been derived from acute cell survival curve equations, modified by consideration of recovery kinetics (Liversage 1969; Barendsen 1982; Thames 1985; Dale 1985; Curtis 1986).

Recent clinical work on low-dose-rate radiotherapy falls into three main areas: interstitial implants, intracavitary treatment and low-dose-rate external beam therapy.

Interstitial implants with radium substitutes employing a range of isotopes with different decay constants have been used in clinical practice (Table 9.3). The dose rate has been adjusted to the range previously used for radium (0.3–1.5 cGy/min) with the exception of that for iodine-125, where rates of 0.1 cGy/min have been used.

For the intracavitary treatment of gynaecological malignancies many centres used to employ radium at a dose rate to point A of approximately 0.5 cGy/min. In order both to improve patient comfort and to increase the number of patients who could be treated, shorter treatment times have subsequently been evaluated using dose rates up to approximately 2 cGy/min. With the development of cobalt-60 intracavitary facilities the opportunity exists for comparison with dose rates of 150–400 cGy/min. However, high-dose-rate treatments are usually employed with different dose-fractionation schedules.

Table 9.3. Exposure rate constants of therapeutic radionuclides

	Half-life	Av. photon energy (MeV)	Exposure rate constant (R $m^2h^{-1}Ci^{-1}$)
Radium-226	1604 years	0.78	0.82
Cobalt-60	5.26 years	1.25	1.3
Caesium-137	30 years	0.66	0.33
Iridium-192	74.2 days	0.35	0.46
Iodine-125	60.2 days	0.027	0.14
Gold-198	2.7 days	0.42	0.24

Use of low-dose-rate external beam therapy for the sparing of lung damage in contrast to haemopoietic tissue makes this approach of critical interest in total body irradiation for leukaemia; this topic is discussed in Chaps. 10 and 11. A study of low-dose-rate local field irradiation in head and neck cancer has been conducted by Pierquin et al. (1978) and this is now being compared with conventional fractionation.

Interstitial Radiotherapy

An isoeffect graph relating largely to the tumour control and normal tissue damage caused by radium implant treatment of oropharyngeal tumours was published by Paterson (1952) "to set out the doses in the 5–10 days range which are equivalent to any desired 7-day dose" (Fig. 9.2). The graph represents clinical experience rather than scientific analysis. The accuracy of the clinical experience was supported by the very similar isoeffect curve published by Ellis (1968) and attributed to observations by Green. These relationships have been used by a number of authors to derive mathematical corrections of dose for isoeffect at different dose rates. For example, Orton (1974) published an isoeffect graph based on various sets of early data from the Christie Hospital in Manchester and used the regression formulae to derive time dose factor (TDF) values for implants at different dose rates.

Some recent clinical reports have failed to support the proposed corrections, but before considering them in detail it is important to emphasize the range of factors influencing the clinical result in interstitial treatments (Table 9.4). These include in particular tumour site, target volume, dose homogeneity and choice of isodose line for reference dose rate (Burgers et al. 1985). Additionally, certain isotopes have a relative biological effect (RBE) which varies with distance from the source – for example iodine-125, where low-energy photons have a high linear energy transfer (LET), and californium-192, where there is a component of neutron irradiation. A further problem of comparisons with iodine-125 is the rapid dose attenuation associated with this isotope and thus the disproportionately reduced integral volume for the same reference dose (Kim and Hilaris 1975).

Table 9.4. Factors affecting implant result

Tumour stage
Treatment volume
Site
Radiation:
Dose
Time
Quality
RBE
Dosimetry
Implant technique

Many of these difficulties were avoided in the retrospective analysis by Pierquin et al. (1973) of the results of iridium-192 implants. The relatively short half-life of the isotope (74 days) meant that patients were sometimes implanted with sources that had partially decayed and thus had a lower linear activity. At equivalent sites and doses, a comparison could be made between dose rates from 70 Gy in 3 days to the same dose in 9 days. Squamous carcinomas of the skin (110 patients), penis (22 patients), lower lip (50 patients) and oropharynx (81 patients) were analysed without detecting a dose-rate effect on tumour control or recurrence. In particular, this was true for the 81 oropharyngeal tumours, where there were 13 local recurrences and 14 instances of late necrosis (Table 9.5), and it was concluded that dose should not be reduced from 70 Gy for shorter treatment times down to 3 days. The Paterson graph recommendation, on the other hand, would have been to reduce Pierquin's standard total dose of 70 Gy in 7 days to approximately 80% (i.e. 56 Gy) for a 3-day treatment and to approximately 90% (63 Gy) for a 5-day implant.

Table 9.5. Iridium implants for tumours of anterior tongue and floor of mouth: impact of dose rate

Dose rate range (cGy/min)	No. of cases	Recurrences	Necrosis
1.2–1.6	9	1	1
0.8–1.2	39	8	9
0.5–0.8	33	4	4

From Pierquin et al. (1973).

An analysis of the effects of dose correction for shorter treatment times has been reported by Awwad et al. (1974) and Awwad and Burgers (1976). Of 26 patients implanted with radium needles for squamous carcinoma of the tongue there were 11 recurrences: 2/7 T_1 tumours, 8/18 T_2 tumours and 1/1 T_3 tumours. When patients remaining recurrence-free and those with recurrence within 2 years were compared, there was no difference in target volume or dose homogeneity between the two groups. However, the group with recurrence had had a significantly shorter treatment time and hence lower dose (Table 9.6). Of the 11 recurrences 9 had treatment times below 110 hours as opposed to only 1 of 15 disease-free patients. This experience has been extended (Burgers et al. 1985) in 119 patients with cancer of the tongue, bladder or perineum. Again the hypothesis that a dose reduction is needed to correct for higher dose rates was not substantiated.

Table 9.6. Tongue implants: clinical results of dose reduction for faster dose rate

	Two years recurrence free	Local recurrence	*p*
No. of patients	15	11	
Av. dose (min. cGy)	6290	5800	0.05
Av. treatment time (hours)	130	100	0.02
No. with treatment time <110 hours	1 (7%)	9 (82%)	

From Awwad and Burgers (1976).

Implantation with iodine-125 illustrates the extreme of low clinical dose rate. Though it is difficult, for reasons discussed above, to interpret clinical results entirely in terms of a dose-rate effect, the protagonists of this treatment modality are in no doubt about the efficacy and lack of normal tissue toxicity of iodine-125. For example, in 112 patients with T_1–T_3 carcinoma of the prostate followed for 24–72 months (median 42 months) after implant 12 (11%) relapsed. There were no bowel complications or instances of urinary incontinence, and no treatment induced impotence (Hilaris et al. 1978). More recent reports have emphasized the relatively high relapse rate in large or poorly differentiated tumours (Giles and Brady 1986). The Memorial Sloan-Kettering Cancer Center has studied iodine-125 implantation for lung tumours and for metastatic cervical lymphadenopathy, comparing results with historical controls implanted with radon or iridium (Kim and Hilaris 1975). Increased local control rates at equivalent levels of complications have been reported.

Dose rates as low as 0.1 cGy/min may be prescribed in iodine-125 implants and the question arises as to whether this might be too low in rapidly proliferating tumours, since studies in vitro suggest that the dose rate required to halt proliferation is a function of the cell cycle time (Hall 1972). For human tumours approximately 0.5 cGy/min might therefore be necessary in some cases. Though patient numbers in the study are small, Gutin et al. (1981) have reported that low-intensity iodine-125 implants (0.1 cGy/min at tumour edge) have failed to control early progression of high-grade glioblastomas, whereas higher activity iodine-125 sources (0.5–1.6 cGy/min) were more successful.

Intracavitary Irradiation

Assessment of the dose-rate effect in the intracavitary treatment of gynaecological malignancies is complicated by the variety of factors that may influence the control rate and the incidence of toxicity (Table 9.7). Data likely to relate to dose rate can only be obtained where applicator design, source distribution, fractionation and ancillary external beam treatment are held stable for different dose rates. With the traditional radium technique (Paterson 1963) the dose rate varies through the pelvis from approximately 1.5 cGy/min at point A to <0.5 cGy/min at the pelvic side wall. Lee et al. (1976) provided suggestive evidence of the importance of dose rate by a detailed retrospective analysis of dose and dose-rate distributions in the pelvis in 15 patients who had suffered complications following radiotherapy for carcinoma of the cervix. In each patient the source distribution was reproduced from radiological records and the maximum dose and dose rate for rectum, bladder and ureter were mapped. In a graphical representation of dose against dose rate for both injured and non-injured organs a line of demarcation could be derived and the slope of this line closely matched that of Paterson's relating to interstitial therapy (see above).

Table 9.7. Factors affecting clinical results of intracavitary irradiation

Applicator design	
Source:	geometry
	activity distribution
	isotope
Afterloading and adaptable source distribution	
Shielding	
Reproducibility and stability	
Prescription	
Total dose	
Dose rate	
Time between treatments	
Fractionation	
External beam:	field
	prescription
	shielding
Patient factors	

Data are also available from centres which have changed their intracavitary sources from radium to either caesium or cobalt. The change to caesium was prompted by considerations of staff exposure to radiation and was usually accompanied by a change to a remote after-loading system. Increased dose rate and hence shorter treatment times were considered to carry benefits of patient comfort, increased unit cost effectiveness and possibly more stable source positioning.

Particularly valuable data come from the Christie Hospital, where early carcinoma of the cervix is treated entirely by intracavitary radiation. The Manchester radium system (Paterson 1963) gave a dose of 7500 cGy at point A in two insertions of 70 hours each over a total treatment time of 10 days. With the change to caesium-137, the dose rate at point A was increased by a factor of approximately 3.5 from *c.* 0.9 cGy/min to *c.* 3.1 cGy/min. In order to assess the appropriate total dose reduction these irradiation schedules were reproduced in an animal model, the assay being based on mouse tail necrosis (Wilkinson et al. 1980). This laboratory study suggested that at 3.1 cGy/min the equivalent normal tissue damage to that produced by treatment at 0.9 cGy/min would require a dose reduction of 33%, that is to a dose to point A of only 50 cGy.

However, in view of the scarcity of clinical data supporting a dose-rate effect, and the broad range of pelvic radiation tolerance, together with a clinical priority of prevention of tumour recurrence over normal tissue damage, it was concluded that a more modest dose adjustment should be evaluated. Following pilot studies, prospective randomized trials were instituted in 1980 to compare traditional radium at a dose of 7500 cGy with caesium-137 at doses of 7500 cGy or 7000 cGy. An early analysis of patients treated up until December 1981 revealed a high incidence of bowel damage in the caesium arms of the trial (Sherrah-Davies 1985) (see Table 9.8). It has been argued that this damage was largely a consequence of technique rather than dose rate, but a further trial was started in 1982 comparing a radium dose of 7500 cGy with caesium-137 at doses of 6500 cGy and 6000 cGy, representing 13% and 20% dose reductions respectively. Preliminary analysis reveals a significantly higher tumour recurrence rate at the 20% dose reduction (R. D. Hunter, personal communication) and it appears that this increased dose-rate after-loading technique is associated with a higher level of normal tissue damage at equivalent control rates when compared with the traditional Manchester radium techniques.

Table 9.8. Incidence of bowel damage in terminal ileum or sigmoid colon following intracavitary irradiation (Christie Hospital 1979–1981)

Isotopic source		Dose (cGy)	No. of patients	% bowel damage
1979	Ra	7500	127	0
Trial:	Ra	7500	33	3
	Cs	7000	30	10
	Cs	7500	26	27

From Sherrah-Davies (1985).

In Glasgow, the corresponding dose-rate change was by a factor of 1.66. It was decided to apply only 50% of the calculated cumulative radiation effect (CRE) correction (Kirk et al. 1972; Corner et al. 1982). Again, early analysis was suggestive of increased normal tissue damage and further study continues with the full CRE correction. These valuable studies will provide data on dose–response both for control of carcinoma of the cervix and for normal tissue damage in the pelvis. However, in both Manchester and Glasgow the associated change in insertion technique and apparatus leads to difficulties in defining the significance of dose rate. In contrast, a current trial at the Institut Gustave Roussy, Paris, is comparing two dose rates in the pre-operative intracavitary irradiation of early carcinoma of the cervix, employing identical technique but two sets of caesium sources. Patients are randomized to received around 60 Gy in either the traditional 6 days or in only 3 days, with no total dose correction. More than 300 patients have been registered in the first 2 years of this study (D. Chassagne, personal communication) and a first report on toxicity will be prepared in 1988.

Extensive studies have been performed on the use of high-dose-rate remote after-loading with cobalt sources in the intracavitary treatment of gynaecological malignancies (Joslin et al. 1972; Bates and Berry 1978; Snelling et al. 1979). Dose rates of approximately 150–300 cGy/min are possible with this system.

Although the dose distribution is similar to that in modern low-dose-rate after-loading technique, the rigidity of the applicators may have a bearing on both the morbidity and effectiveness of treatment when compared with traditional radium techniques.

Dose equivalence calculations have been derived from acute cell survival formulae (Liversage 1969) and from the isoeffect curve of Ellis (Joslin et al. 1972). Although the total dose has been reduced in line with these calculations, clinical experience has resulted in fractionation schedules being developed in most centres that give 4–6 fractions of 600–900 cGy at either weekly or twice-weekly intervals. This treatment has almost universally been combined with external beam regimens employing some type of central blocking. It is consequently difficult to estimate equivalent doses of high-dose-rate and low-dose-rate treatments. A prospective trial of high-dose-rate treatment versus either low-dose-rate after-loading or conventional radium therapy has not been performed.

Calculations of the CRE for two different regimens of fractionated high-dose-rate treatment combined with external beam treatment showed close correlation of these values with both local tumour control and morbidity (Joslin and Sharma 1984). These results suggested that small changes in the high-dose-rate regimen could have considerable effects on the results of treatment.

External Beam Irradiation

The use of low-dose-rate external beam radiotherapy is derived mainly from the requirement for long source–skin distance (SSD) tele-cobalt treatments to achieve large field sizes for whole body irradiation. This is discussed in Chaps. 10 and 11. More limited investigations have also been undertaken on skin irradiation and on the treatment of head and neck cancer.

The influence of dose rate on skin damage was studied in patients following surgery for breast cancer when adjuvant irradiation fields included the parasternal areas (Turesson et al. 1984). Each parasternal field was treated separately at different dose rates to a total dose of 7.89 Gy × 4 over 22 days (one fraction per week). The right side was treated within 4 min per fraction (≥2 Gy/min) and the left side was treated over 32 min in three fractions of 2.63 Gy with 14–15-min intervals between fractions (average approximately 0.25 Gy/min). The lower dose rate reduced both acute skin erythema and late telangiectasia, the required dose correction for isoeffect being approximately 1.1 (Turesson et al. 1984).

The exploration of low-dose-rate external beam irradiation for radical treatment has been restricted for practical reasons. A pilot study in 46 patients with advanced squamous tumours of the head and neck was reported by Pierquin et al. (1978). Patients were treated using a tele-cobalt apparatus with a weak 6-mm cobalt source such that 1.7 cGy/min would be achieved with an 8 cm × 8 cm field at 6 cm depth with a SSD of 80 cm. Treatment times were 6–9 hours per day with brief rests every 2 hours and a 30–45-minute lunch break. The couch incorporated a mattress and pillow and immobilization in a supine position was assisted by sandbags only.

A total dose of 6500–7000 cGy was reached either by single course treatment over 9–16 days or by a split course over 5 weeks. Thirteen patients were treated by single course treatment with a high local control rate. However, acute reactions were severe and most patients suffered extensive mucositis for 5–6 weeks. Three suffered moist desquamation of skin and seven developed late necrosis. The split course approach was adopted subsequently with a 3-week gap after the first 35 Gy in most cases, and of 21 patients with at least 1 year follow-up the local control rate was similar to that in the single course treatment with considerably less toxicity. Seven patients (33%) had late necrosis, especially severe when larynx or piriform sinus were in the field.

This approach is now being compared with conventional fractionation in a prospective randomized trial in patients with T_{2B} and T_{3A} squamous carcinomas of the oropharynx (Pierquin et al. 1985). The low-dose-rate split course technique (7000 cGy over 5 weeks) is compared with conventional fractionation at 200 cGy per day (7000 cGy over 7–8 weeks). With either schedule the final 2500 cGy could be delivered by interstitial implant. The randomization was not rigorous because of reservations about the low-dose-rate arm. Twenty-nine of the 53 patients were treated by conventional fractionation and there was a disproportionate number of T_2 tumours in this group. Nevertheless there were fewer local recurrences in the low-dose-rate group (26% at 2 years versus 53% for conventional fractionation) (Fig. 9.3). All five recurrences after low-dose-rate treatment occurred at the lower end of the dose-rate spectrum (2/2 at 1.5 cGy/min, 3/14 at 2.25 cGy/min), whereas all three patients treated at $\geq$2.8 cGy/min developed necrosis.

This use of fractionated low-dose-rate irradiation at 750 cGy/day is similar in toxicity to schedules employing multiple fractions per day in order to accelerate treatment (Gonzalez et al. 1980). Doses of 4800–5400 cGy were given over the first 11–14 days via three fractions per day of 1.8 Gy mid-plane dose. It is not clear, however, whether the two approaches are equivalent with respect to tumour control or late effects.

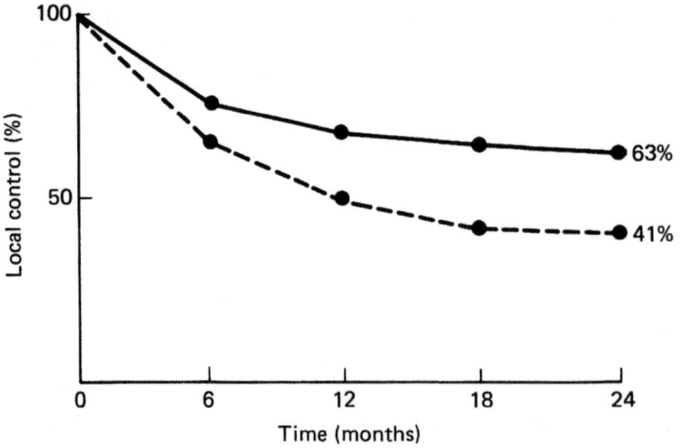

Fig. 9.3. Local control of oropharyngeal tumours treated by conventional fractionation (*dotted line*) or by low-dose-rate (*continuous line*) external beam therapy. (After Pierquin et al. 1985.)

Conclusions

It is now almost 10 years since Hall's editorial (1978) asked "Has the promise of low dose rate irradiation been realized?" The promise derived historically from the excellent clinical results of interstitial implants, and Hall felt that the major causes for this were that:

1. The distribution of the radiation dose was highly localized.
2. The continuous radiation may have benefits relating to cell cycle distribution, leading to cells accumulating in the more radiosensitive phases of the cell cycle.
3. The dose is delivered in a short treatment time.

Additionally, low-dose-rate radiation may exploit a difference in repair capacity between tumour and critical normal tissue.

In clinical situations these factors are often interrelated and the relative importance of each may vary between different tissues and different tumours. For example, it is likely that the small target volume contributes to tolerance of rapid high-dose treatment in interstitial implant therapy. However, implants with iodine-125 dissect these factors since treatment time is so protracted with this isotope. Though it does not appear that interstitial therapy with iodine-125 is inferior to that with other isotopes, comparison is difficult because of dose and RBE differences.

Pierquin's studies on low-dose-rate external beam therapy are an attempt to exploit the possible benefits of continuous irradiation. There is also the additional rationale quoted of the possibility of a low oxygen enhancement ratio compared with conventionally fractionated irradiation (Hall et al. 1966; Pierquin et al. 1978). The initial study also incorporated a very short overall duration of treatment; this proved too toxic however. The current study also employs a shorter than conventional overall treatment time in the low-dose-rate arm (7000 cGy in 5 weeks versus 7000 cGy in 7–8 weeks) (Pierquin et al. 1985). Most current investigations into shortening of the overall treatment time are based on accelerated fractionation studies rather than low dose rate.

Gynaecological studies have explored the opposite direction from these considerations on the basis that increased patient comfort and staff safety should compensate for any disadvantages of a slight increase in the dose rate for intracavitary treatments. Following dose correction for normal tissue effects it will remain to be seen whether dose-rate changes (or duration of insertion) are important for tumour control.

Total body irradiation seeks to exploit the repair differences between critical normal tissues and leukaemic cells (see Chap. 11). Low-dose-rate sparing of lung tissue has been demonstrated both experimentally and clinically. However this can also be accomplished by fractionation and thus a specific advantage for continuous low-dose-rate therapy in this situation has yet to be demonstrated.

References

Awwad HK, Burgers JMV (1976) Studies in dose-time volume relationships in bladder and tongue radium implants. Clin Radiol 27:443–448

Awwad HK, Burgers JMV, Marcuse HR (1974) The influence of tumour dose specification on the early clinical results of interstitial tongue implants. Radiology 110:177–182

Barendsen GW (1982) Dose fractionation, dose rate and iso-effect relationships for normal tissue response. Int J Radiat Oncol Biol Phys 9:1981–1997

Bates TD, Berry RG (1978) High dose-rate after-loading in the treatment of cancer of the uterus. Br J Radiol [Special Report]: 17

Burgers JMV, Awwad HK, Van der Laarse R (1985) Relation between local cure and dose-time volume factors in interstitial implants. Int J Radiat Oncol Biol Phys 11:715–723

Corner GA, Kirk J, Perry AM (1982) Low dose-rate afterloading techniques: choice of dose and time. Clin Radiol 33:145–147

Curtis SB (1986) Lethal and potentially lethal lesions induced by radiation – a unified repair model. Radiat Res 106:252–270

Dale RG (1985) The application of the linear-quadratic dose–effect equation to fractionated and protracted radiotherapy. Br J Radiol 58:515–528

Ellis F (1968) The relationship of biological effect to dose-time-fractionation factors in radiotherapy. Curr Top Radiat Res 4:357–397

Giles GM, Brady LW (1986) 125-Iodine implantation after lymphadenectomy in early carcinoma of the prostate. Int J Radiat Oncol Biol Phys 12:2117–2125

Gonzalez D, Breur K, Van der Schueren E (1980) Preliminary results in advanced head and neck cancer with radiotherapy by multiple fractions a day. Clin Radiol 30:417–421

Gutin PH, Phillips TL, Hosobuchi Y et al. (1981) Permanent and removable implants for the brachytherapy of brain tumours. Int J Radiat Oncol Biol Phys 7:1371–1381

Hall EJ (1972) Radiation dose-rate: a factor of importance in radiobiology and radiotherapy. Br J Radiol 45:81–97

Hall EJ (1978) The promise of low dose rate: has it been realized? Int J Radiat Oncol Biol Phys 4:749–750 (editorial)

Hall EJ, Bedford JS, Oliver R (1966) Extreme hypoxia: its effect on survival of mammalian cells irradiated at high and low dose-rates. Br J Radiol 39:302–307

Hilaris BS, Whitmore WF, Batata MA, Barzell W, Tokita N (1978) ^{125}I implantation of the prostate: dose–response considerations. Front Radiat Ther Oncol 12: 82–90

Joslin CA, Sharma SC (1984) Brachytherapy 1984. Nucleotron Trading BV, The Netherlands

Joslin CAF, Smith CW, Mallik A (1972) The treatment of cervix cancer using high activity ^{60}Co sources. Br J Radiol 45:257–270

Kim J, Hilaris B (1975) Iodine-125 source in interstitial tumor therapy. Am J Roentgenol Radium Ther Nuc Med 123:163–169

Kirk J, Gray WM, Watson ER (1972) Cumulative radiation effect. II. Continuous radiation therapy–long-lived sources. Clin Radiol 23:93–105

Lee KH, Kagan AR, Nussbaum H, Wollin M, Winkley JH, Norman A (1976) Analysis of dose, dose-rate and treatment time in the production of injuries by radium treatment for cancer of the uterine cervix. Br J Radiol 49:430–440

Liversage WE (1969) A general formula for equating protracted and acute regimes of radiation. Br J Radiol 42:432–440

McWhirter R (1936) 13th annual report of the British Empire Cancer Campaign. London, pp 131–144

Mitchell JS (1960) Studies in radiotherapeutics. Blackwell, Oxford

Orton CG (1974) Time-dose factors (TDFs) in brachytherapy. Br J Radiol 47:603–607

Orton CG, Ellis F (1973) A simplification in the use of the NSD concept in practical radiotherapy. Br J Radiol 46:529–537

Paterson R (1952) Studies in optimum dosage. Br J Radiol 25:505–516

Paterson R (1963) The treatment of malignant disease by radiotherapy, 2nd edn. Williams and Wilkins, Baltimore

Pierquin B, Chassagne D, Baillet F, Paine CH (1973) Clinical observations on the time factors in interstitial radiotherapy using iridium-192. Clin Radiol 24:506–509

Pierquin BM, Mueller WK, Baillet F (1978) Low dose rate irradiation of advanced head and neck cancers: present status. Int J Radiat Oncol Biol Phys 4:565–572

Pierquin B, Calitchi E, Mazeron JJ, Le Bourgeois JP, Leung S (1985) A comparison between low dose rate radiotherapy and conventionally fractionated irradiation in moderately extensive cancers of the oropharynx. Int J Radiat Oncol Biol Phys 11:431–439

Regaud C (1922) Distribution chronologique rationelle d'un traitement de cancer epithelial par les radiations. Compt Rend Soc de Biol 86:1085–1088

Sherrah-Davies E (1985) Morbidity following low-dose-rate electron therapy for cervical cancer. Clin Radiol 36:131–139

Snelling MD, Lambert HE, Yarnold JR (1979) The treatment of carcinoma of the cervix and endometrium using the cathetron at the Middlesex Hospital. Clin Radiol 30:253–258

Thames HD (1985) An "incomplete-repair" model for survival after fractionated and continuous irradiations. Int J Radiat Biol 47:319–339

Turesson I, Notter G, Wickstrom I, Johansson K-A, Eklund S (1984) The influence of irradiation time per treatment session on acute and late skin reactions: a study on human skin. Radiother Oncol 2:235–245

Wilkinson JM, Hendry JH, Hunter RD (1980) Dose-rate considerations in the introduction of low-dose-rate afterloading intracavitary techniques for radiotherapy. Br J Radiol 53:890–893

10 Total Body Irradiation: Normal Tissue Effects

L. S. Constine and P. Rubin

"The whole is greater than the sum total of its parts."

Introduction

Irradiation to the total body in high dose occurs unintentionally and with potential life-threatening consequences following nuclear reactor accidents and nuclear warfare, and intentionally as a therapy for certain malignancies. Total body irradiation (TBI) in these settings, and with a low dose and dose rate as an incidental exposure during space travel, has stimulated a renewed interest in the biological effects of TBI. The lethal syndromes that follow high-intensity irradiation are related to the doses which the whole of specific sensitive organs receive (Fajardo 1982). Different threshold dose levels exist for irreversible injury to the stem cell populations of various organs or tissues. Once exceeded the result is loss of the organ's functional integrity coupled with an inability to repopulate or repair (Hall 1978). The degree of importance of the organ determines the survival of the individual, and the time course of the expression of injury relates to the cell kinetics and cycle time of the stem cell. The classic radiation lethality syndromes relating to bone marrow, gastrointestinal and central nervous system failure each have a tolerance dose and time course to fatality (Rubin and Casarett 1968). A brief review of the critical dose factors and the clinical-pathological course of events associated with each of the radiation syndromes will constitute the first part of this chapter.

The therapeutic application of TBI was stimulated by the ability to transplant bone marrow and thus protect the most radiosensitive and vulnerable organ system. The combination of chemotherapy and TBI led to resistant or recurrent childhood leukaemias being cured (Thomas et al. 1982). With this as a model, the concept of supralethal radiation treatment with bone marrow rescue is currently being investigated for a variety of radiosensitive disseminated malignancies such as non-Hodgkin's lymphoma, Ewing's sarcoma, small cell anaplastic lung cancer and neuroblastoma (O'Reilly 1983; Appelbaum and

Buckner 1986). The most frequent lethal toxicity which has appeared following bone marrow reconstitution is interstitial pneumonitis (Meyers et al. 1982). The various possible aetiologies for this complication include infection (frequently with the cytomegalovirus), graft-versus-host disease and radiation/chemical pneumonopathy which may also interact with and exacerbate the other causes. Other lethal syndromes resulting from radiation injury will become apparent as rescue of more sensitive or earlier-reacting organs is accomplished (Sullivan et al. 1984).

TBI syndromes should be viewed as a constellation of lethal syndromes each resulting when the radiation tolerance of a whole organ, exposed by necessity in the course of TBI, is exceeded. To recognize the potential late effects that can occur, it is necessary to have an understanding of the dose–response data for each whole organ, in both the adult and the child, since organ radiosensitivity is greater during the rapid growth phase (Rubin et al. 1982). Radiation oncologists have long recognized that the tolerance of an organ to radiation is improved by shielding or protecting a part of its volume. Thus, whole organ radiation data are sparse. Furthermore, improved strategies for administering TBI are rapidly evolving with respect to both dose rate and fraction size. Thus continuous single radiation exposures which previously were given at a high dose rate (>0.25–0.5 Gy/min) are now more commonly delivered at a low rate (<0.1 Gy/min) (Evans 1983). More recently fractionated and hyperfractionated whole body radiation regimens have been used (Shank 1983). For some malignancies with cells exquisitely sensitive to irradiation, such as acute lymphatic leukaemia (ALL) and non-Hodgkin's lymphoma, very small daily fractions of TBI (0.05–0.15 Gy to total doses of 0.5–1.5 Gy) have surprisingly exceeded bone marrow tolerance. This suggests that a spectrum of responses can occur depending on the specific radiation schedule used (Rubin et al. 1981).

Each organ system will be reviewed here, and available data presented on the sensitivity to injury from single continuous doses and fractionated regimens in both clinical and experimental settings, with the goal of determining whole organ dose–response curves. Equally important is the documentation of the effect of chemotherapy, particularly in combination with irradiation, but this is beyond the scope of this chapter.

Lethal Radiation Syndromes

When the whole body is irradiated intensively with single exposures, death will occur within minutes to months as a function of dose and the organ system affected. Damage to other organs, not directly leading to death, will additionally contribute qualitatively or quantitatively to the syndrome. The prominent acute syndromes and modes of death after single-dose TBI are, in the order of occurrence and decreasing threshold dose, the central nervous system (CNS) syndrome, the gastrointestinal (GI) syndrome and the haemopoietic syndrome (Rubin and Casarett 1968). Table 10.1 summarizes a variety of parameters of the acute radiation syndromes in man.

Animal data available on these syndromes are generally consistent with observations in man. For example, haemopoietic death occurs in 50% of humans

Table 10.1. Some aspects of the acute radiation syndromes in man (after whole body irradiation)

Aspects	Acute syndromes in whole body irradiation		
	Central nervous system (CNS) syndrome	Gastrointestinal (GI) syndrome	Haemopoietic syndrome
Chief determining organ	Brain	Small intestine	Bone marrow
Syndrome threshold	20 Gy	5 Gy	1 Gy
Syndrome latency	¼–3 hours	3–5 days	2–3 weeks
Death threshold	50 Gy	10 Gy	2 Gy
Death time	Within 2 days	3 days to 2 weeks	3 weeks–2 months
Characteristic signs and symptoms	Lethargy, tremors, convulsions, ataxia	Malaise, anorexia, nausea, vomiting, diarrhoea, GI malfunction, fever, dehydration, electrolyte loss, circulatory collapse	Malaise, fever, dyspnoea on exertion, fatigue, leukopenia, thrombopenia purpura
Major underlying pathology	Vasculitis (CNS), encephalitis, meningitis, oedema (CNS)	Depletion of intestinal epithelium, neutropenia (marrow damage), infection	Bone marrow atrophy, pancytopenia, infection, haemorrhage, anaemia

From Rubin and Casarett (1968).

in 30–60 days following single doses of 2.4–7.5 Gy (Bond et al. 1965). Similarly, the haemopoietic $LD_{50/30-60}$ is 6–7 Gy in rodents, 6 Gy in rabbits, 2.5 Gy in pigs and goats, and 6 Gy in the macaque monkey (Bond et al. 1965). For any particular organ-related syndrome the actual threshold dose may not be clinically relevant because the adverse effects resulting from injury to another organ system may predominate. Thus, when the dose is at the threshold for the CNS syndrome, it is far above the threshold for the GI and haemopoietic syndromes. Clear differences exist in the timing of the development of the pathology which leads to functional impairment from the loss of relevant cells in the determining organs. Thus, the CNS syndrome becomes apparent within a few minutes to hours after irradiation and continues to express itself during the early parts of the latent periods for the GI and haemopoietic syndromes. Similarly, when the dose permits survival of the individual through the period of the CNS syndrome but exceeds the threshold for the GI syndrome, then this syndrome becomes apparent during the latent period for the haemopoietic syndrome. It should be noted that the median acute lethal dose ($LD_{50/60}$) for man for brief, intensive TBI is not precisely known, nor is the influence of age or sex. However, this dose has been estimated to be between 3 and 5 Gy, with deaths primarily associated with the haemopoietic and GI syndromes.

Survival after TBI intentionally administered as preparation for bone marrow transplantation (BMT) results from the infusion of viable bone marrow and intensive supportive care. Patients so treated will demonstrate delayed adverse effects dependent on the tolerance of individual organs to irradiation of their entirety. Since TBI for BMT is currently administered in both single and fractionated doses, the tolerance of whole organs to these two ablative strategies must be considered.

Whole Organ Tolerance: Single Versus Fractionated Doses

Whole organ tolerance is determined by the radiosensitivity of the relevant stem cell subpopulations, which may not always be proliferating or dividing (Hall 1978). The functional capacity of cells is often distinct from the regenerative capacity and permits organ physiology to be preserved in the face of injury and allows for recovery or repair of the insult. Most organ systems are composed of many cell subpopulations, that is, 20 to 40 or more, each performing an important activity (Rubin and Casarett 1968). The most radiosensitive vital cell population determines organ tolerance and organ failure just as the degree of importance of an organ that has been irradiated determines the survival of an organism.

Acute radiation effects can lead to severe pathophysiological organ failure and are different from the late effects; both need to be considered. It is well recognized that late effects can occur without being preceded by acute effects (Rubin 1977; Siemann et al. 1982). The stem cells or target cells responsible for these late effects are usually slowly proliferating or virtually non-proliferating except if stimulated by injury or loss of cells – for example liver hepatocytes after surgical resection. Of equal, if not greater importance is the microvasculature of tissue in general (Rubin and Casarett 1968).

As noted, the radiation response is related to dose and the time in which it is delivered, in terms of both fraction number and dose rate. Sufficient clinical and experimental studies exist to provide a profile of the whole organ tolerance for the acute and late effects of TBI. Organ tolerance to single and fractionated doses needs more complete and systematic study and documentation before the results can be considered firm data. However, such a listing can be generated using available data, as in Table 10.2.

Table 10.2. Tolerance radiation doses for whole organs: single and fractionated

Single dose (Gy): $T_{5/5}$–$T_{50/5}$	
Bone marrow 2–10	Heart 18–20
Lens 2–10	Liver 15–20
Lung 7–10	Mucosa 15–20
Thyroid 7.5	Skin 12–20
GI tract 10–20	Testes >20
Kidney 10–20	Spinal cord 20–25
Ovary >10–40	Brain 20–30
Fractionated dose (Gy): $TD_{5/5}$–$TD_{50/5}$	
Testes 1.5–2.5	Liver 35–40
Ovary 5–15	Mucosa 30–40
Lens 6–20	Skin 30–40
Bone marrow 15–30	Heart 40–50
Kidney 23–28	GI tract 45–50
Lung 25–30	Spinal cord 50–60
Thyroid 30–40	Brain 60–70

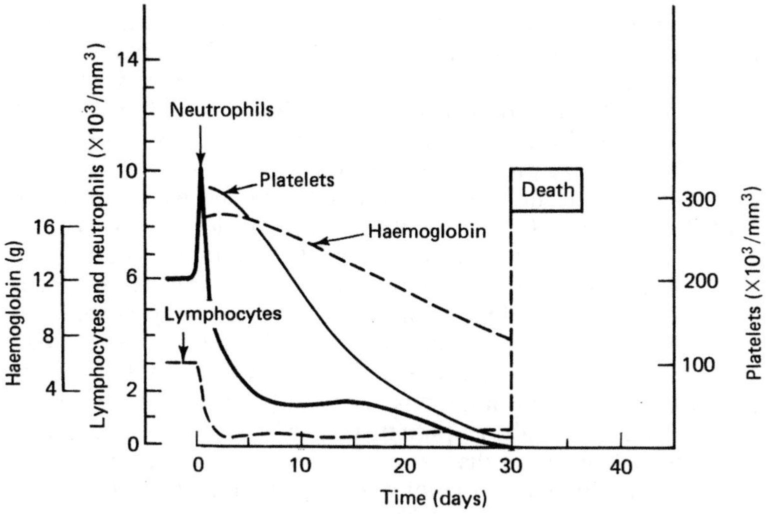

Fig. 10.1. Expected haematological response of a human following a single dose, total body exposure of 4.5 Gy. (After Andrews 1967.)

Bone Marrow

Single Doses to Whole Volume

As described in the preceding section on TBI syndromes, bone marrow is one of the most radiosensitive organs, responding to doses of 1.5–7.5 Gy with a rapid depletion of vital stem cells within a week of exposure (Tubiana et al. 1979). Death usually occurs as a result of granulocytopenia and thrombocytopenia predisposing the individual to overwhelming infection and hemorrhage (Fig. 10.1) (Andrews 1967). The recent reactor safety estimates for an $LD_{50/30}$ are at variance with information from nuclear warfare estimates. Doses of 3–5 Gy result in an $LD_{50/100}$, with a more frequent death rate (and shorter interval to death) resulting from 7.5–10.5 Gy (Reactor Safety Study 1975). Following bone marrow transplants, doses of 7.5–10.5 Gy are well tolerated and the microvasculature of the marrow still allows for the implantation and proliferation of transferred stem cells (Kim et al. 1980).

Fractionated Doses to Whole Volume

Fractionated doses to the whole bone marrow organ are rarely used in clinical schedules except for treatment of certain leukaemias, multiple myelomas and non-Hodgkin's lymphomas, and more recently in the form of TBI for BMT (Bergsagel 1971; Carbell et al. 1979; Jaffe et al. 1979; Thomas et al. 1982). In CLL extremely small daily doses of 0.05–0.15 Gy were effectively used to total doses of 0.5–1.5 Gy, with the total regimen at times administered more than once (Rubin et al. 1981). In non-Hodgkin's lymphoma these higher dose schedules were more frequently well-tolerated, with 0.10–0.15 Gy being given twice a week to a total of 0.3 Gy weekly for 5 weeks to 1.5 Gy. Cycles of treatment were given monthly and could lead to total doses as high as 3.0–4.5 Gy; however, severe to life-threatening haematological toxicity sometimes occurred (Carbell et al. 1979). A process of titration of dose is essential since severe thrombocytopenia can occur, and treatment should be withheld if the platelet count falls below 100 000. Patients with multiple myeloma and CLL do appear to exhibit more haematological sensitivity to TBI than those with normal bone marrow and death has resulted after very modest radiation doses (Bergsagel 1971; Rubin et al. 1981). Data from animal models suggest that this sensitivity is atypical. A series of experiments in rabbits documented that small daily doses of 0.05, 0.10 or 0.25 Gy given repeatedly to very high total doses are well tolerated, greatly exceeding the single dose $LD_{50/30}$ (Rubin et al. 1984). To explain the sensitivity seen in the above patient groups it has been postulated that the haematopoietic cells in these diseases do not respond appropriately to an irradiation stimulus by an increase in colony stimulating factor in the serum, as leukopenia occurs (Scarantino et al. 1984).

The suppressive effects of irradiation on bone marrow are volume dependent and are of considerable importance clinically. Tolerance levels appear greater if permanent aplasia of limited areas is the end-point assessed. Knopse has reported that single doses in excess of 20 Gy are required whereas fractionated regimens produce aplasia only with doses greater than 30 Gy (Knopse et al. 1968). Rubin showed that even higher doses of 40 Gy are required to suppress later regeneration at 2–5 years (Rubin 1973). The volume of bone marrow irradiated has a major impact on the regeneration pattern (see Table 10.3) (Rubin 1984).

Table 10.3. Bone marrow regeneration (BMR) patterns and compensatory mechanisms

Techniques of irradiation	Regeneration			Doses (Gy)	
	Exposed bone marrow	Unexposed bone marrow	Extension	Daily	Total
Small field	N	Local-regional ↑ BMR	N	2	>40
Large field	N	Generalized ↑ ↑ BMR	N	2	>30
Subtotal body	Suppressed BMR which then recovers	Generalized ↑ ↑ BMR	↑ ↑	2	40
Total body	Active	—	N	0.05–0.1	>1

Reprinted with permission from *International Journal of Radiation Oncology Biology Physics*, vol. 10. P. Rubin, The Franz Buschke Lecture. © 1984 Pergamon Journals Ltd.
N, none; ↑, increased activity; ↑ ↑, greatly increased activity.

Following the loss of bone marrow activity in 10%–25% of the bone marrow organ after irradiation to this volume, the unexposed bone marrow responds by increasing its population of progenitor cells (Rubin 1984). Often the bone marrow organ fails to regenerate the ablated portion of marrow since the compensatory process is able to meet the demands for haematopoiesis. However, when a larger volume of bone marrow is irradiated (approaching 50%) the paradoxical phenomenon of in-field regeneration is seen 2–5 years later, as well as extension of bone marrow into previously quiescent long bones within 1–2 years. This is well illustrated by the evaluation and mapping of bone marrow with nuclear scans using ^{99m}Tc-colloid undertaken in Hodgkin's disease treated with different radiation field arrangements (Rubin et al. 1973).

Lung

Single Doses to Whole Volume

Single dose radiation exposures to the lung are common as an incidental component of TBI programmes widely used in preparatory regimens for BMT. They are also employed in patients with widespread metastases. The relatively high radiosensitivity of the lung was appreciated when patients who had survived hemibody regimens suffered fatal interstitial pneumonitis with sufficient frequency that investigators realized the event was not a manifestation of lymphatic carcinomatosis involving lung (Fryer et al. 1978). Through careful analysis, Keane et al. (1981) determined exact dose correction factors. That is, supervoltage and in particular telecobalt radiotherapy displayed an increased lung transmission of 15%–20%. The dose range for lethal radiation-induced pneumonitis is firmly established at 8.2 Gy for a 5%, 9.3 Gy for a 50% and 11.00 Gy for an 80% incidence following a single high-dose-rate exposure (Fig. 10.2) (Fryer et al. 1978; Keane et al. 1981). This dose–response relationship is shifted to the right for protracted low-dose-rate irradiation as used in TBI for BMT

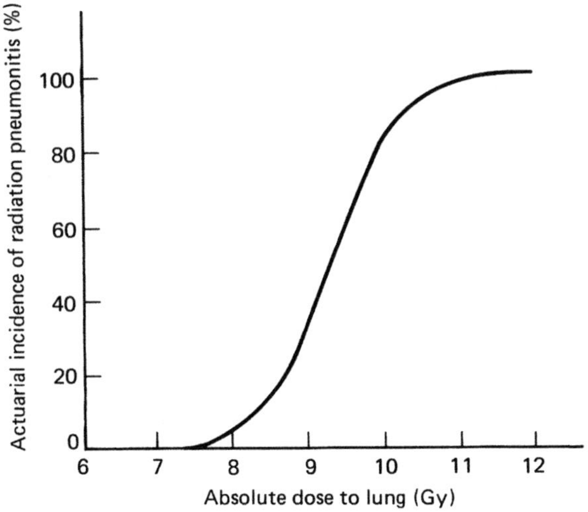

Fig. 10.2. A best-fit curve based on a probit regression analysis of actuarial incidence of radiation pneumonitis versus absolute dose to lung for patients receiving upper hemibody irradiation. (Reprinted with permission from *International Journal of Radiation Oncology Biology Physics*, vol. 7. T. J. Keane, J. Van Dyk and W. D. Rider, Idiopathic interstitial pneumonia following bone marrow transplantation. © 1981 Pergamon Journals Ltd.)

(Keane et al. 1981). In mice, the single-dose regimen curves are also very steep, in the range of 2 Gy for a change from an LD_{10} to an LD_{90} (Siemann et al. 1982).

Fractionated Doses to Whole Volume

Fractionated irradiation dramatically improves tolerance in both experimental and clinical settings. Wara and Phillips demonstrated that in a mouse model the $LD_{50/160}$ changed from 13.65 Gy in a single fraction to 38.20 Gy in ten fractions for a dose modification factor of 2.79 (Table 10.4) (Wara et al. 1973). Also important is the effect of dose rate, as considerable protection is conferred by lowering the rate from 0.47 to 0.08 Gy/min (Evans 1983). The $LD_{50/30}$ thus changes from 7.75 Gy to 8.70 Gy for a dose modification ratio of 1.12 (Evans 1983). The relevance of dose rate in several animal species has been summarized recently by Lockhart (Table 10.5) (Lockhart et al. 1986). The clinical data for fractionated regimens have been carefully reconstructed by Phillips and Wara,

Table 10.4. LAF_1 mouse lung $LD_{50/160}$

No. of fractions	Days	$LD_{50/160}$ (95% confidence limits)	NSD[a]	ED[b]
1	—	13.24 (12.81–13.65)		
2	1	16.98 (16.25–17.69)	14.38	13.08
4	3	23.18 (22.68–23.67)	14.73	12.90
10	11	36.94 (35.72–38.20)	16.33	13.49
10	25	38.73 (37.44–39.97)	15.64	13.49
20	25	46.68 (45.56–47.95)	15.96	12.52

From Wara et al. (1973).
[a] From the formula: Total dose = $NSD \cdot N^{0.24} \cdot T^{0.11}$.
[b] From the formula: Total dose = $ED \cdot N^{0.377} \cdot T^{0.058}$.

Table 10.5. Sparing afforded by low-dose-rate thoracic irradiation

Species	Dose rate (cGy/min) High	Dose rate (cGy/min) Low	DRF[a]	Assay, time of evaluation
Mice	110	5	1.8	Lethality, 26 weeks
Rats	80	5	1.7	Lethality, 26 weeks
Mice	100	6	1.8	Lethality, 26 weeks
Mice	100	5	1.7	Breathing rate, 14 weeks
			1.6	Lethality, 26 weeks
Humans	50–400		1.3	Clinically diagnosed idiopathic interstitial pneumonitis
Humans		2.8–15		

Reprinted with permission from *International Journal of Radiation Oncology Biology Physics*, vol. 12. S. P. Lockhart, J. D. Down and G. G. Steel, The effect of low dose rate and cyclophosphamide on the radiation tolerance of the mouse lung. © 1986 Pergamon Journals Ltd.
[a] DRF, dose reduction factor.

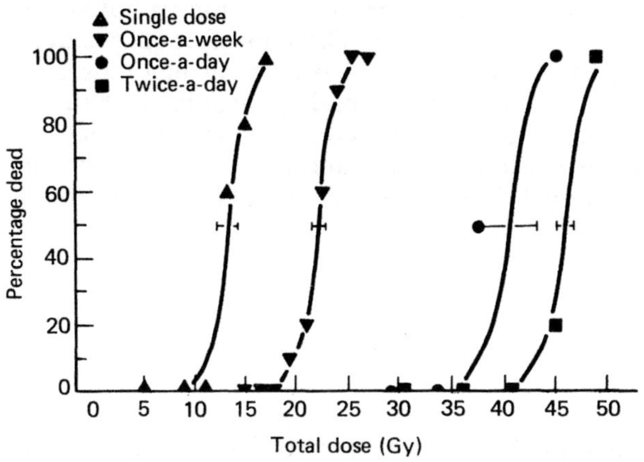

Fig. 10.3. Lethality dose–response curves for single or fractionated radiation exposures (28 weeks after a single dose).

particularly as regards children with Wilms' tumour, many of whom received whole lung irradiation with or without actinomycin. The dose for a 5% lethality ($TD_{5/5}$) was 26.50 Gy in 20 fractions, or 770 rets, and for 50% lethality was 30.50 Gy in 20 fractions, or 992 rets. The dose modification factor for single versus fractionated doses is 3.2, demonstrating a considerable increase in tolerance for the latter (Wara et al. 1973).

Studies in our laboratories by Siemann illustrate the effect of increasing time intervals and decreasing fraction sizes on lung injury after irradiation (Fig. 10.3). Considerable protection is evident when comparison is made between single exposures and once-a-week, once-a-day or twice-a-day fractionation regimens. When the end-point for evaluation is not only acute pneumonitis but the late appearance of fibrosis, a similar alteration in tolerance dose–response data is again noted (Salazar et al. 1984). The appearance of the late effect without an antecedent acute effect illustrates an important principle: the need to follow animals for long intervals in order truly to appreciate the late effects, much as in patients. The dose to produce late-appearing interstitial septal fibrosis is less than that for early pneumonitis independent of the fractionation scheme used (Rubin et al. 1984).

Doses to Partial Volume

The adverse effects of single versus fractionated doses to partial lung volumes have not been well studied experimentally or fully documented clinically. Although different experiences are recorded as regards the incidence of pneumonopathy in patients with lung or breast cancer, a specific range of injurious doses has not been identified. To complicate this problem, patients with lung cancer also have pulmonary abnormalities which are the result of changes produced by the cancer itself (Choi et al. 1985). Assessment with computerized tomographic (CT) scans has more recently provided sharper end-points for the recognition of fibrosis. The dose for the induction of fibrosis is 40 Gy and increases to 50 Gy for pneumonopathy. The impact of split-course and once-a-week irradiation schedules on the incidence of pneumonitis and fibrosis has been shown by Salazar, and confirms the findings in animals (Salazar et al. 1984). That is, the incidence of fibrosis is increased relative to that of pneumonitis if patients survive for several months, and the degree of fibrosis produced is a function of the fractional dose as well as the total dose and total time for delivery.

Central Nervous System

The relative tolerance of the mature brain to moderate doses of irradiation would suggest that this organ could sustain the range of therapeutic doses used in TBI without significant adverse effects, and that malignancies within its substance could be adequately treated. However, many patients receiving TBI have had previous courses of CNS irradiation (as, for example, in ALL). Also, primary CNS cancers require high doses of irradiation.

The variety of sequelae which follow CNS irradiation can be categorized according to their time of appearance and histopathology (Sheline et al. 1980). Acute reactions infrequently appear after 50 Gy in fractionated doses and are generally mild with malaise and nausea as symptoms. They are probably due to a transient cerebral oedema and respond to corticosteroids. Early delayed reactions, manifested by somnolence and a variety of neurological signs, occur within 2–3 months in approximately 20% of patients and presumably result from a transient demyelinization (Hoffman et al. 1979). Symptoms are usually mild, although they can occasionally be severe and result in death (Rider 1963). Late delayed reactions include subacute leukomyelopathy in the spinal cord, and leukoencephalopathy, mineralizing microangiopathy or frank necrosis of the white or grey matter in the brain (Price 1979; Sheline et al. 1980).

Extensive fractionation studies have been performed on the tolerance of the rodent spinal cord, with a value of about 0.4 for the N exponent and essentially zero for the T exponent in the NSD isoeffect formula (Fig. 10.4) (Thomas et al. 1975; Horsey and White 1980). Leith recently reviewed available data on the dose–response relationship of radiation-induced spinal cord paralysis in the rat, and found that the estimated dose needed to produce paralysis in 50% of animals ranged from 19 Gy to 25.6 Gy (Van der Kogel 1977; Leith et al. 1981). Fractionated doses, however, produce paralysis over an extremely large total dose range. Ang has estimated the ED_{50} in rats to be as high as 140.00 Gy in 2.00 Gy fractions (Ang et al. 1985). Thus far, the applicability of such animal findings to man in unclear.

The tolerance of the whole brain to single doses of irradiation was studied by Kemper in the monkey (Kemper et al. 1977). Essentially no effect was seen after 10 Gy; a scatter of focal lesions was present after 15 Gy at 26 weeks, with a confluence of lesions at 52 weeks; and extensive necrosis with ventricular enlargement and lethality occurred after 20 Gy. The frequency of brain necrosis following single irradiation doses was also studied in rabbits, and found to be strictly dependent on the volume of brain exposed. Necrosis occurred after 21–

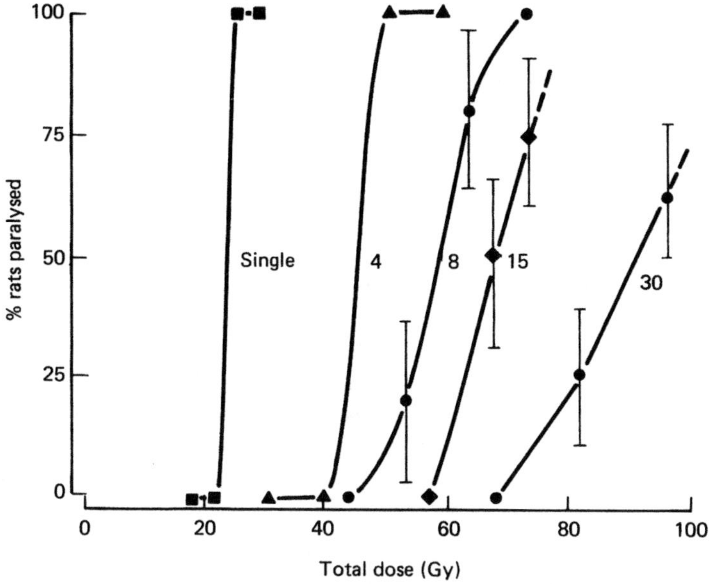

Fig. 10.4. Probability of paralysis in rats after irradiation of a segment of the spinal cord with either a single dose or various numbers of fractions, all given in 6 weeks. (After Horsey and White 1980.)

50 Gy to maximal volumes (Fig. 10.5) (Berg and Lindgren 1963). Clinical data indicate that 45–55 Gy to the brain and 50 Gy to the spinal cord will result in a 5% frequency of severe complications in 5 years, with 75% of cases occurring within 3 years (Sheline et al. 1980). Data for single dose irradiation are sparse, but when Hindo (Hindo et al. 1970) used single doses of 10 Gy in patients with brain metastases, 7% of patients died within 48 hours. Following 10 Gy TBI in preparation for BMT, Thomas (Thomas et al. 1976) found a 7% incidence of leukoencephalopathy at 1–5 months in patients who were previously treated with 12 doses of 2 Gy for ALL prophylaxis.

Heart

Acute and chronic pericarditis are well-recognized complications of irradiation incidentally administered to the heart in the course of treating malignancies in the thorax (Byhardt et al. 1975; Stewart and Fajardo 1984). Other manifestations of chronic damage are also becoming increasingly apparent with the passage of time, but fortunately are infrequent. Pancarditis results from diffuse ischaemia and fibrosis of the muscle layers, and functional valvular injury, predominantly affecting the mitral, aortic and tricuspid valves, can occur (Catterall and Evans 1960; Cohen et al. 1967; Brosius et al. 1981). Conduction defects arise due to fibrosis of the atrioventricular node and conducting branches (Cohen et al. 1967; Brosius et al. 1981; Cohen et al. 1981). Finally, coronary

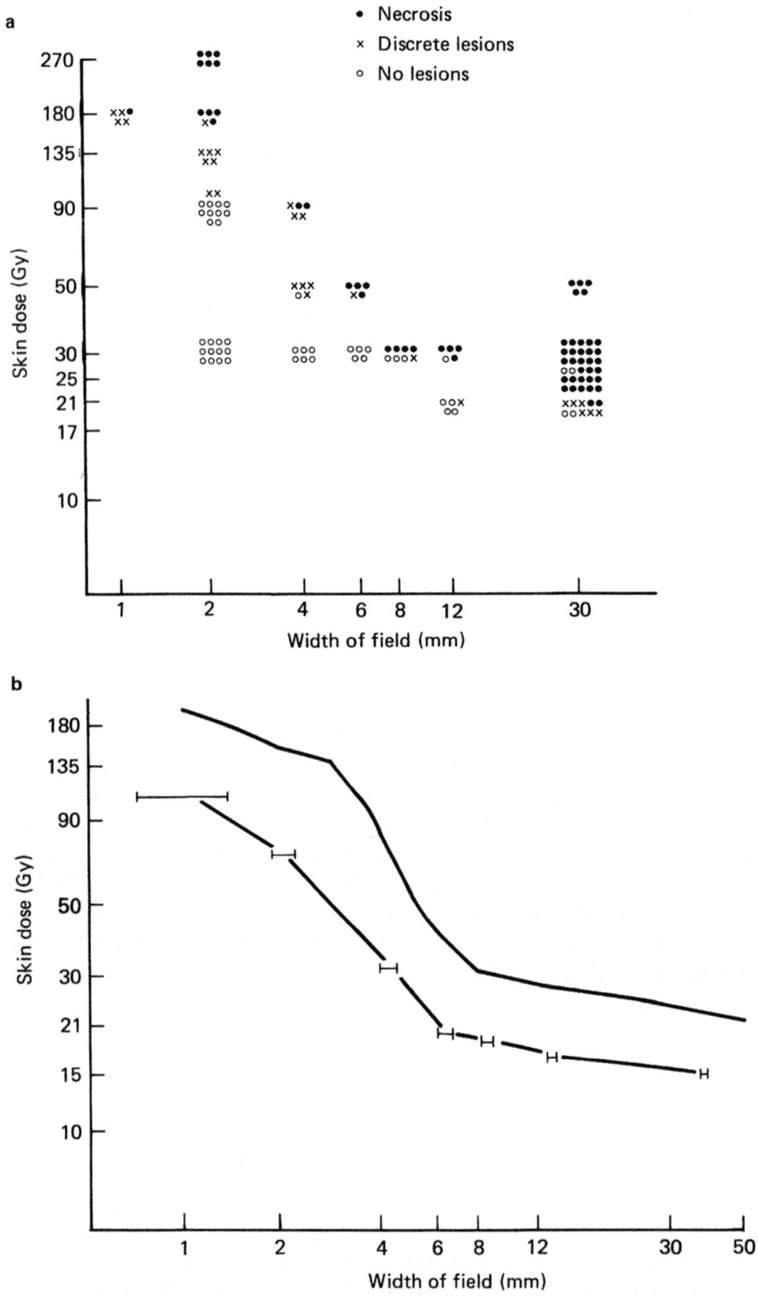

Fig. 10.5. **a** Results of morphological analysis of radiation lesions in the rabbit brain in relation to the skin dose and the width of field, plotted in a log–log diagram. **b** Correlation between width of field, radiation dose and radiation injury in the rabbit brain. *Upper curve*: "tolerance index" based on frequency of lesions after irradiation with a given (skin) dose through a given width of slit. *Lower curve*: lowest focal dose giving morphological lesions in relation to "effective width" of field for different groups of animals. (After Berg and Lindgren 1963.)

artery disease, usually affecting the left anterior descending artery, can result in heart attacks at a young age (Annest et al. 1983; Dunsmore et al. 1986). Pericarditis is well described (Byhardt et al. 1975; Stewart and Fajardo 1984) and 10%–15% of affected patients develop progressive effusive-constrictive disease. Furthermore, delayed pericarditis 5–10 years following radiation therapy even in the absence of acute disease may occur (Stewart and Fajardo 1984).

Experimental studies in a rabbit model were performed by Stewart, who found that single doses of 18–20 Gy or fractionated doses of 54 Gy (4.5 Gy ×12) caused pericarditis in all animals, while less than 18 Gy (single dose) caused no damage (Fajardo and Stewart 1973). Pathologically the following was observed: between 6 and 48 hours following irradiation an inflammatory exudate was seen in all layers of the heart, and after 70 days progressive pericardial and myocardial fibrosis leading to tamponade and heart failure occurred in 87% of animals (Fajardo and Stewart 1970). The time course of the evolution of these lesions is shown in Fig. 10.6 (Stewart and Fajardo 1984). Gavin studied beagles and showed that single doses of 12.2 Gy resulted in pericardial effusion while 15.0 Gy caused tamponade in 50% of animals at 6 months (Gavin and Gillette 1982).

In patient populations, the relevance of dose to subsequent cardiac disease is underscored by differences in the incidence of disease following different radiation techniques which deliver more or less than 40 Gy to the heart. Thus, among patients treated for Hodgkin's disease, a 5% incidence of pericarditis occurs when less than 40 Gy is administered through equally weighted anterior and posterior portals, whereas a 30% incidence is seen following anteriorly weighted techniques (Byhardt et al. 1975). With such techniques, Applefeld

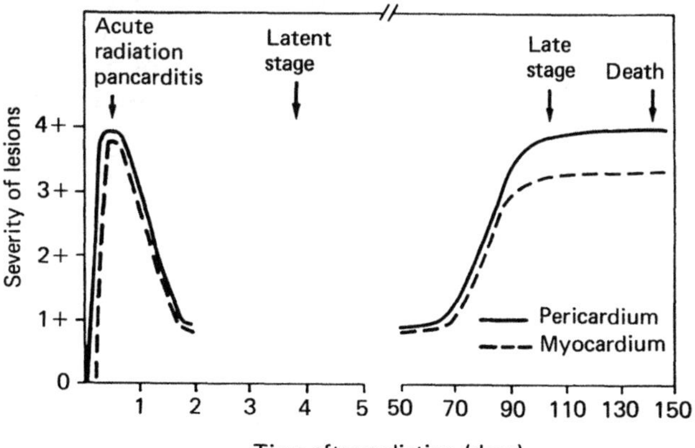

Fig. 10.6. Development of radiation heart disease in the rabbit as observed sequentially by light microscopy after a single dose of 20 Gy given on day 0. The curves indicate the variation in severity of the pericardial and myocardial lesions, but the nature of the lesions differs through the course of the disease: during the first 2 days there is a transient acute exudate of granulocytes in all tissues of the heart, including the valves. No lesions are seen (by light microscopy) between 3 and 70 days. Beyond 70 days there is progressive irreversible fibrosis of the pericardium and myocardium. Important ultrastructural damage of the microcirculation does occur, though, in the myocardium between 3 and 70 days. (After Stewart and Fajardo 1984.)

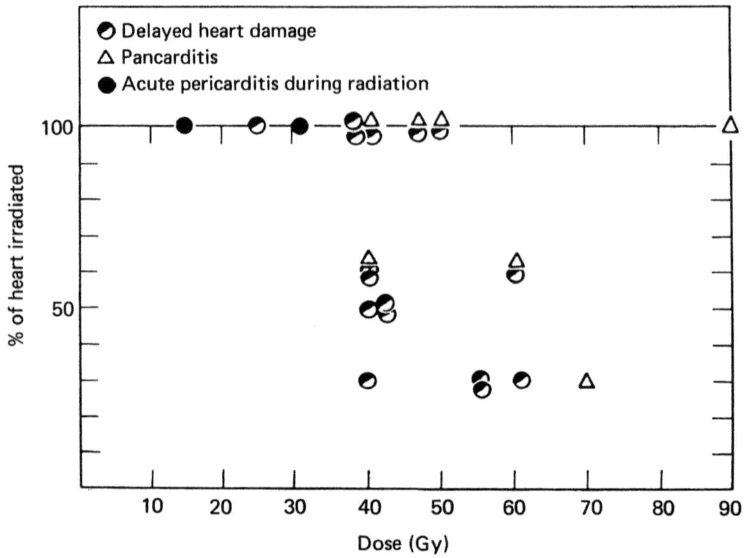

Fig. 10.7. The effect of different doses of radiation to a sizable volume of the heart. After a dose of 10 Gy pericardial reactions are noted. Pericarditis tends to occur with higher doses of 50–70 Gy. (After Stewart et al. 1967.)

found a 64% frequency of clinically apparent or occult pericarditis, a 12% frequency of coronary artery disease, and chronic pericarditis in 11% of cases (Appelfeld et al. 1981). Brosius described autopsy findings in patients with Hodgkin's disease of pericardial thickening in 94%, myocardial fibrosis in 50%, fibrotic mural endocardial thickening in 75%, valvular thickening in 81% and coronary lesions in more than 75% following a calculated dose of 55.92 Gy anteriorly (39.32 Gy posteriorly) (Brosius et al. 1981). The relationship of the volume of the heart irradiated and dose administered to the occurrence of cardiac damage is suggested by early data collected by Stewart (Fig. 10.7) (Stewart et al. 1967).

Thyroid Gland

Thyroid dysfunction may result from direct irradiation injury to this gland. Patients treated for Hodgkin's disease with mantle irradiation are known to develop primary thyroid dysfunction. This is manifested by elevated serum thyrotropic hormone (TSH) with or without a concomitant decrease in thyroxine (T_4) values; these abnormalities occur with greater frequency as the fractionated dose to the whole gland increases from 15 Gy to 50 Gy (Glatstein et al. 1971; Constine et al. 1984) (Fig. 10.8). Different series have reported the frequency of elevated TSH to range from 4% to 88% (Glatstein et al. 1971; Nelson et al. 1978; Devney et al. 1984). Uncompensated hypothyroidism (decreased T_4 and

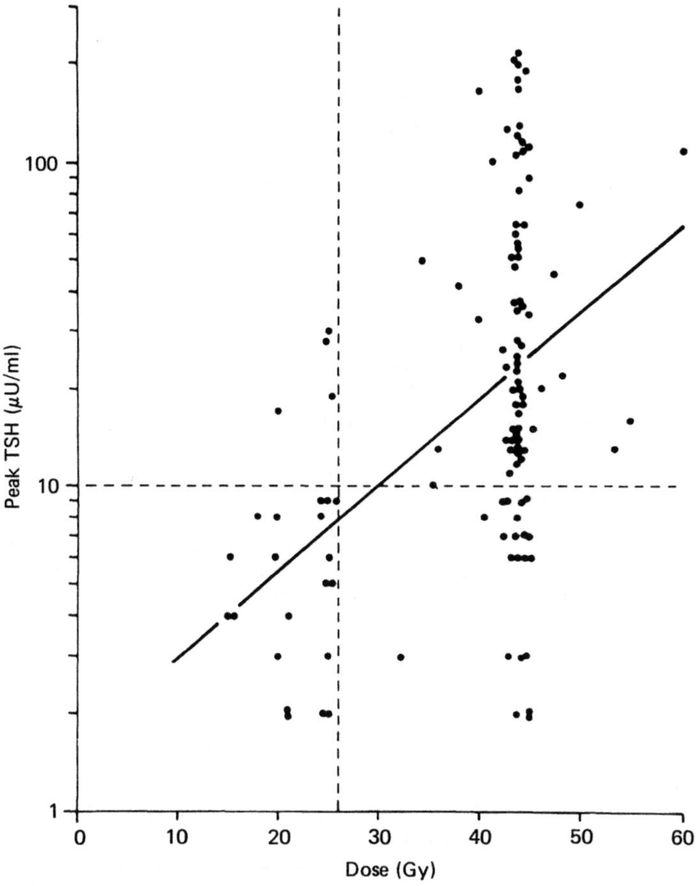

Fig. 10.8. Peak TSH test value plotted against radiation dose, excepting one patient with hyperthyroidism and two euthyroid patients with thyroid nodules. The slope of the regression is 2.7/Gy and is significantly different from zero ($P \leq 0.001$). The corresponding doubling dose for the geometric-mean peak TSH is 11 Gy. (After Constine et al. 1984.)

elevated TSH) is observed to occur in 6%–20% of patients. Data on hypothyroidism resulting from single doses of irradiation are provided by Sklar, who found that 7.5 Gy administered as TBI in preparation for BMT caused a decrease in T_4 in 9% of patients, and an elevated TSH in 35% (Sklar et al. 1982).

Kidney

The sensitivity of the kidney to radiation-induced injury is such that these organs can limit the delivery of optimal doses to tumours in their vicinity. Moreover, careful observation for long-term untoward effects will be necessary for patients

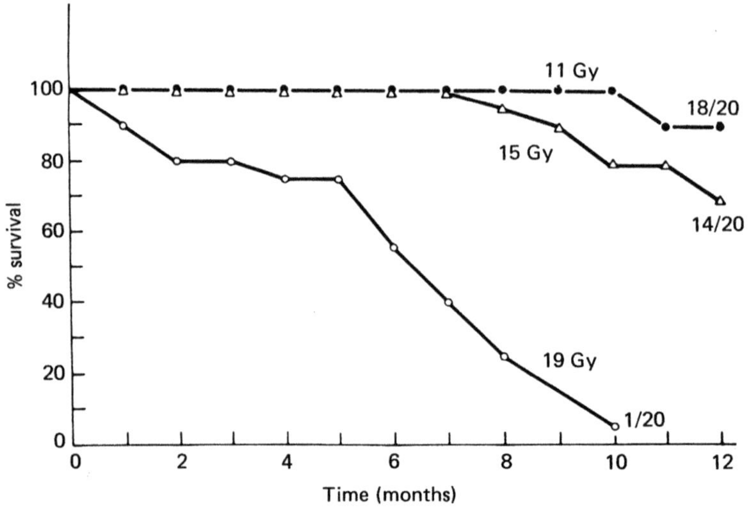

Fig. 10.9. Survival of mice as a function of time following single doses of external radiation to both kidneys. (Reprinted with permission from *International Journal of Radiation Oncology Biology Physics*, vol. 2. E. Glatstein, L. F. Fajardo and J. M. Brown, Radiation injury in the mouse kidney: I. © 1977 Pergamon Journals Ltd.)

receiving TBI in preparation for BMT. The spectrum of pathophysiological changes which result from radiation to the kidney includes a progression from acute nephritis to chronic nephritis, benign hypertension, malignant hypertension, proteinuria and frank renal failure (Maier 1972).

Following TBI, nephrosclerotic changes occur consistent with progressive arteriolonephrosclerosis due to degeneration of arterioles or capillaries (Wachholz and Casarett 1970). Glatstein performed experiments in mice showing that single doses of greater than 19 Gy to both kidneys caused renal failure and death, whereas 11 Gy allowed a 90% survival (Glatstein et al. 1977) (Fig. 10.9). Jongejan showed in rats that in the months following single doses of up to 15 Gy to both kidneys the glomerular filtration rates and urine osmolality progressively deteriorated, and the systolic blood pressure rose, though recovery was observed after lower doses (Jongejan et al. 1987).

In humans, dose–response data following high single dose exposure to both kidneys are scarce. Fractionated doses of 10–20 Gy cause a decrease in the glomerular filtration rate, a 38%–87% reduction in renal plasma flow, and suppression of tubular excretory capacity up to 12 months following therapy; the blood urea and maximum urinary concentrating ability, however, remain normal after these doses (Maier 1972). If the whole of a single kidney is treated with fractionated doses of greater than 26 Gy, a 35% decrease in creatinine clearance is seen at 5 years (Table 10.6) (Willett et al. 1986). Overall, the sensitivity of the kidney to failure following fractionated irradiation was calculated to be 23 Gy for a 5% frequency at 5 years, and 28 Gy for a 50% frequency at 5 years. This suggests that a slight increase in dose could cause significant additional damage after a threshold has been exceeded (Laxton and Kunkler 1964).

Table 10.6. Percentage decrease in mean creatinine clearance according to percentage of kidney irradiated (>26 Gy) and time after radiotherapy

Year	50% kidney irradiated		60%–85% kidney irradiated		90%–100% kidney irradiated	
	No. of observations	% decrease in creatinine clearance	No. of observations	% decrease in creatinine clearance	No. of observations	% decrease in creatinine clearance
1	8	11 (±26)	6	14 (±9)	5	26 (±33)
2	18	9 (±32)	4	12 (±18)	12	21 (±45)
3	12	21 (±28)	5	22 (±54)	5	23 (±10)
4	1	—	1	—	6	14 (±16)
5	3	3 (±44)	2	53 (±18)	6	35 (±13)
>5	6	7 (±36)	3	24 (±17)	7	29 (±18)

Reprinted with permission from *International Journal of Radiation Oncology Biology Physics*, vol. 12. C. Willett, J. Tepper, E. Orlow et al., Renal complications secondary to radiation treatment of upper abdominal malignancies. © 1986 Pergamon Journals Ltd.
Numbers in parentheses indicate standard deviation.

Gastrointestinal Tract

The GI tract includes many organs, which demonstrate a spectrum of sensitivities to radiation-induced injury. The terminal ileum is most frequently symptomatically damaged, as a result of the high turnover rate of epithelial cells (cell replacement in the crypts and villi occurs every 3–6 days) (Hall 1978). Following single doses of 5 Gy to greater than 15 Gy, patients experience nausea, vomiting and diarrhoea leading to dehydration (Maier 1972). Pathological changes include a cessation of mitosis, crypt cell pyknosis, fragmentation, and swelling and vacuolation of the cells in the enteric mucosa. After 6–8 hours mucosal cells demonstrated a transient proliferation with a burst of atypical mitoses, and over the next 48 hours cell loss without renewal is progressive, with shortening of the crypts and villi. Subsequently the villi show progressive denudation resulting in a loss of protein and electrolytes. Following lower radiation doses recovery with a chronic reaction may ensue, with the submucosa most severely affected. Collagen and bizarre fibroblasts replace fatty tissue and vascular lesions occur. Delayed effects can take 10 years or more to develop (Hall 1978).

Trott (1984) studied changes in the rat rectum following single doses of irradiation and found that doses greater than 20 Gy caused severe proctitis in 50% of animals within 40–200 days; 24.5 Gy in two fractions caused similar changes. Hubmann (1981) showed that a single dose of 22.5 Gy induced structures in the small intestine in 50% of rodents at 100 days, and 21.5 Gy caused the same effect in the rectum. Rectal obstruction increased in frequency from 0 to 100% with increasing single doses of from 15 to 30 Gy.

In man, single doses of as little as 1.5–3 Gy will cause acute effects with necrosis in the walls of crypts (Fajardo 1982). The tolerance of the different gastrointestinal organs to fractionated irradiation is estimated to be as shown in Table 10.7 (Roswitt et al. 1972).

Table 10.7. The tolerance of gastrointestinal organs to fractionated irradiation

Organ	$TD_{5/5}$ (Gy)	$TD_{50/5}$ (Gy)
Oesophagus	60	75
Stomach	45	50
Intestine	45	60
Colon	45	65
Rectum	55	80

Testis

Testicular dysfunction in the form of azoospermia or hormonal alterations results from extremely low doses of irradiation. It may thus occur as a consequence of the exposure to scattered irradiation delivered during therapy for such highly curable conditions as Hodgkin's disease, as well as after TBI in preparatory regimens for BMT (Speiser et al. 1973; Sklar et al. 1984). This tissue sensitivity is convincingly illustrated by Heller (1967), who documented changes in sperm counts in man after various single irradiation doses (Byhardt et al. 1975) (also see Fig. 10.10). The sensitivity of the sperm is dependent on the stage of development of this cell lineage. Lushbaugh (Lushbaugh and Casarett 1976) reviewed animal data and noted that mouse spermatogonia die after single doses

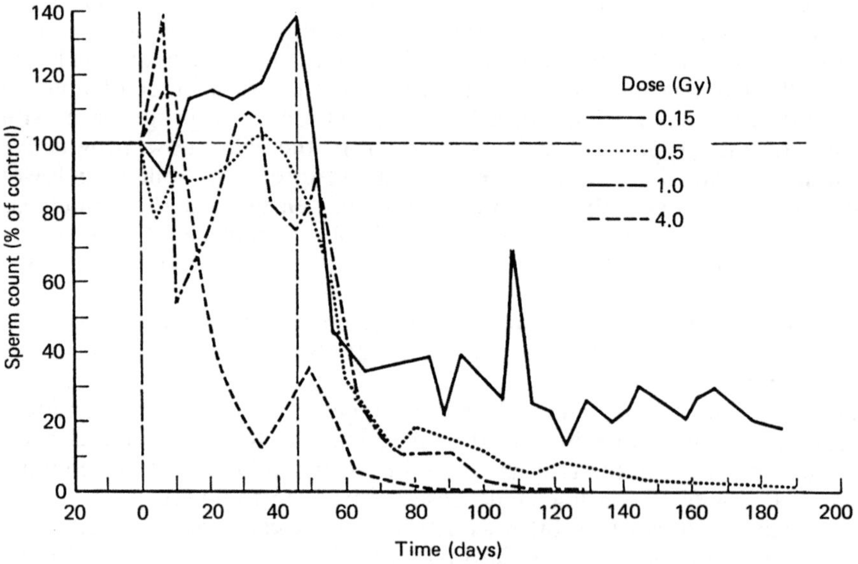

Fig. 10.10. Sperm counts of normal men following various single high-intensity X-ray (190 kVp) exposures to testes. (After Lushbaugh and Casarett 1976.)

of less than 0.35 Gy, while 2–10 Gy is necessary to kill spermatocytes, and 15 Gy to ablate the spermatid population. The sensitivity of this cell line is species dependent, as emphasized by the observation that more than 600 Gy is needed to kill mature rabbit spermatids. Moreover, the time to recovery is also different in different species, with rodents recovering more quickly than man. However, with azoospermia as an end-point, the sensitivity of different animal species following single and fractionated radiation doses is broadly similar, as suggested by the data in Table 10.8 (Speiser et al. 1973).

Table 10.8. The doses of single or fractionated radiation required to produce azoospermia in different animal species

Species	Single dose (Gy)	Fractionated dose (Gy)
Rabbit	15–25	20–23
Rat	10–30	8–10
Mouse	16–30	5–7.5
Guinea-pig	45	21

In man, a differential sensitivity of sperm according to the stage of development is also seen, with single doses of 0.15 Gy killing spermatogonia, 2 Gy killing spermatocytes, and 5–6 Gy being necessary to kill mature spermatids (Lushbaugh and Casarett 1976). Because of the kinetics of cell maturation, spermatogonia in man and animals are more sensitive to fractionated radiation doses, with azoospermia occurring after 2–3 Gy of scattered irradiation to the testes in the course of treating Hodgkin's disease. Recovery may take 3–14 years (Pedrick and Hoppe 1986). In fact, doses as small as 0.28–1.35 Gy over 4–5 weeks can cause temporary azoospermia; recovery from these lower doses is more rapid, taking 1–3 years (Pedrick and Hoppe 1986). After single doses of irradiation in man the overall (albeit somewhat different) sensitivities are as described in Table 10.9.

Table 10.9. The sensitivities of sperm production to single doses of radiation

Reference	$TD_{5/5}$	$TD_{50/5}$
Rubin and Casarett (1968)	1.5–2.5 Gy	20 Gy
Lushbaugh and Casarett (1976)	<4.5 Gy	>6 Gy

Hormonal alterations after therapeutic irradiation are also well described. Shapiro (Shapiro et al. 1985) has documented decreases in follicle stimulating hormone at 6 months, with recovery over 30 months, following scattered irradiation doses of 1–25 Gy in 25–30 fractions; full recovery occurred only if the total dose was less than 0.5 Gy. Full recovery of luteinizing hormone production occurred at doses of less than 2 Gy. Again, different observers have noted different sensitivities. Thus Shalet has noted that Leydig cell function was not impaired following 2.68–9.83 Gy (20 fractions over 4 weeks) but that failure did occur after 24 Gy (over 16–22 days) (Shalet et al. 1978, 1985).

Ovary

When girls and adult females are irradiated, the static populations of mature ovarian oocytes do not repopulate after destruction (Lushbaugh and Rider 1972). This may explain why the dose of irradiation necessary to obliterate all the oocytes is larger in younger than in older women. For women over the age of 40 years, 6 Gy in fractionated doses can cause permanent menopause (Heller 1967; Lushbaugh and Casarett 1976), while 50% of younger women will become sterile after 2 Gy (Fig. 10.11) (Heller 1967; Lushbaugh and Rider 1972). Overall, a 5% incidence of sterility is seen at 5 years after 2–6 Gy, and a 50% incidence following doses greater than 6–20 Gy, depending on patient age (Doll and Smith 1968; Lushbaugh and Casarett 1976). Following 10 Gy in a single dose, in the form of TBI for BMT, prepubertal girls showed an absence of menses and development of secondary sexual characteristics; pubertal women all experienced ovarian failure and 50% had menopausal symptoms (Sklar et al. 1983).

Animal data are again species dependent and related to the stage of oocyte maturation. In the mouse, the least mature follicles are sensitive to less than 0.1 Gy, while more mature follicles are ablated after single doses of less than 1 Gy or fractionated doses of less than 3–4 Gy (Lushbaugh and Casarett 1976). In the dog, single doses of greater than 20 Gy are necessary, and fractionated doses (0.3 Gy/day) of 4.75 Gy (Lushbaugh and Rider 1972).

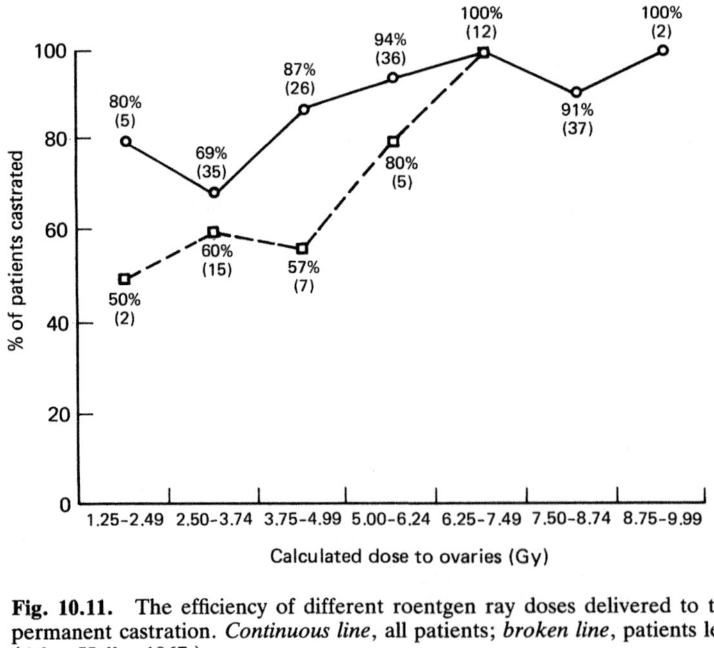

Fig. 10.11. The efficiency of different roentgen ray doses delivered to the ovaries in producing permanent castration. *Continuous line*, all patients; *broken line*, patients less than 40 years of age. (After Heller 1967.)

Eye

Visual impairment resulting from cataract formation is a side-effect which is clearly not life-threatening but yet extremely distressing for patients. Abnormalities originate in the posterior pole of the lens and subsequently progress at a pace dependent on the radiation exposure (Merriam et al. 1972). Merriam documented a 100% incidence of cataracts when more than 2 Gy were delivered to the lens in a single dose, but a total dose greater than 11.5 Gy was necessary for that frequency of cataract formation when the radiation was fractionated (Merriam and Focht 1957). A 60% frequency of cataracts (50% of which were progressive) occurred after 7.5–9.5 Gy in fractionated doses, and only rare cataracts if less than 4 Gy were administered (Fig. 10.12). The interval to abnormality was generally 2–3 years, although the range was 6 months to 35 years (for higher or lower doses respectively) and averaged about 8 years after 2.5–6.5 Gy. More recently Deeg has recorded the frequency of cataracts following TBI. A single dose of 10 Gy causes cataracts in 80% of patients, while 12–15 Gy in fractionated doses results in only a 19% frequency.

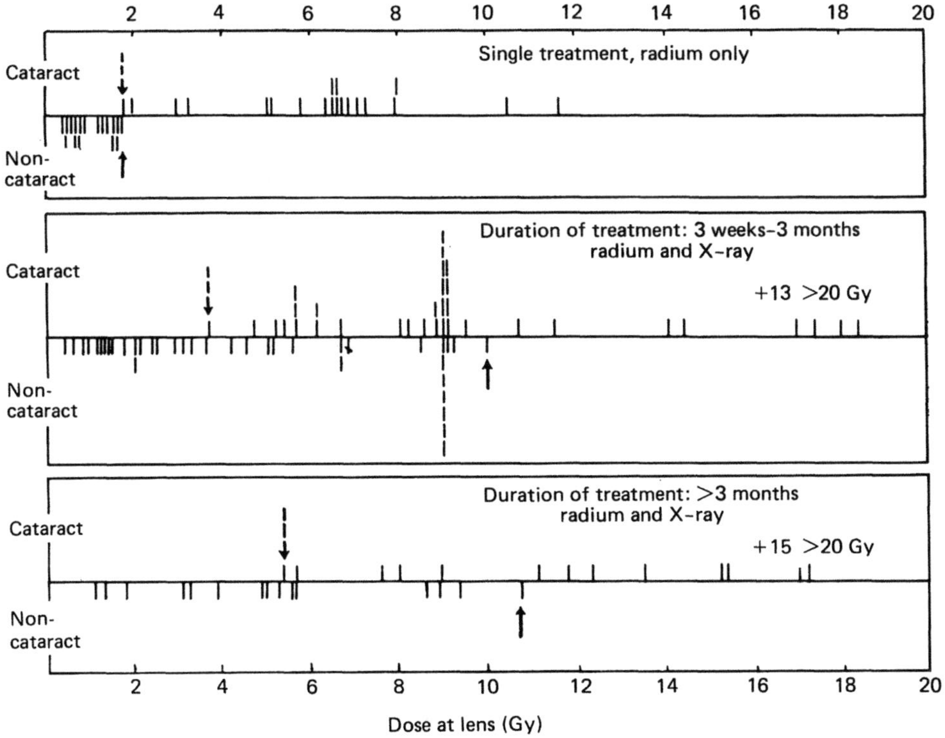

Fig. 10.12. Doses of X- or gamma-radiation to lens in 97 cases of radiation cataract and 70 cases without lens opacities. (After Merriam et al. 1957.)

Liver

The relative sensitivity of the liver to irradiation injury precludes the eradication of infiltrating tumours from this organ by high doses to its entirety. Following such doses a series of pathological changes occur including hyperemia, an increase in volume, dilatation and congestion of sinusoids, atrophy of hepatocytes, and veno-occlusive lesions appearing as early as 2.5–6 months (Rubin and Casarett 1968). The patient develops ascites, and there is an increase in hepatic size and a rise in bilirubin and alkaline phosphatase levels. Ingold (Ingold et al. 1965) reviewed 40 patients who received total liver irradiation in the course of therapy for ovarian carcinoma or lymphoma, and found that no cases of radiation-induced liver disease occurred below a dose of 25 Gy, whereas 21% of patients had abnormalities following a dose of 30–36 Gy, and 42% following 38–42 Gy (Fig. 10.13). Tefft studied changes in hepatic function in children older than 5 years of age. He observed that fractionated doses of less than 25 Gy caused abnormal liver function tests and radionuclide scans in approximately 50% of patients, while 25–35 Gy caused abnormalities in 63%, and greater than 35 Gy was highly toxic to 86%. In adults, Phillips (Phillips et al. 1984) showed that substantial hepatic damage occurred at fractionated doses of

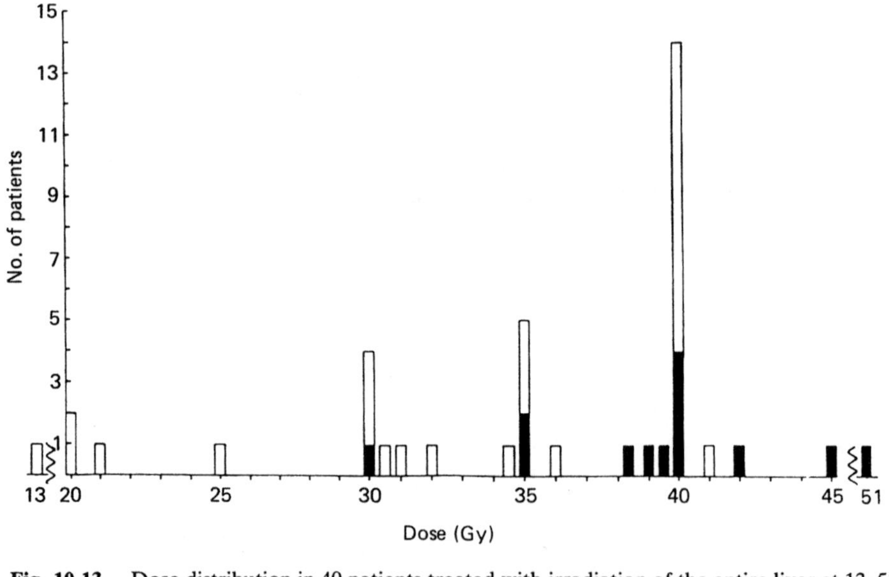

Fig. 10.13. Dose distribution in 40 patients treated with irradiation of the entire liver at 13–51 Gy. The incidence of radiation hepatitis appears to be related to the dose of hepatic radiation given. *Open columns*, total number of patients; *filled columns*, cases of radiation hepatitis. (After Ingold et al. 1965.)

greater than 20–37.5 Gy and Kim (Kim et al. 1976) described similar changes after more than 30 Gy. Veno-occlusive disease is an uncommon but severe complication of TBI administered in preparation for BMT, and can occur following single doses as low as 7.5 Gy (Woods et al. 1980). Short intense fractionation schedules, such as are used in abdominal strip field techniques for the treatment of ovarian cancer, lower the threshold dose to 15–20 Gy (Perez et al. 1978).

Conclusions

The above discussion of whole organ tolerance to single dose and fractionated irradiation indicates that several organs not commonly presumed to be at risk for chronic injury following TBI for BMT will, in fact, demonstrate effects if the patient survives. This will be more common if early-reacting vital organs such as bone marrow and lung are salvaged through such measures as marrow replacement, innovative irradiation schedules, etc. The organs which subsequently demonstrate adverse effects will do so in order of their chemosensitivities and radiosensitivities. Thus the lens, gonads and kidney will be expected to suffer ill-effects, as has already been noted by several observers. The thyroid, liver, heart and other organs are also likely to demonstrate abnormalities of significance with the passage of time. Such effects, as in the case of the heart, might cause substantial morbidity and even death.

It is of interest to consider the relative potential for radiation damage following fractionated as compared with single dose radiation schedules. It will be important for the radiation oncologist to monitor these organs carefully, using scoring systems for late effects, and to devise strategies to minimize the inevitable morbidity.

References

Andrews GA (1967) Radiation accidents and their management. Radiat Res [Suppl] 7:390–397

Ang K, Van der Kogel A, Van der Schueren E (1985) Lack of evidence for increased tolerance of rat cord with decreasing fraction doses below 2 Gy. Int J Radiat Oncol Biol Phys 11:105–110

Annest LS, Anderson, RP, Wei-i L et al. (1983) Coronary artery disease following mediastinal radiation therapy. J Thorac Cardiovasc Surg 85:257–263

Appelbaum FR, Buckner CD (1986) Overview of the clinical relevance of autologous bone marrow transplantation. Clin Haematol 15:1–18

Applefeld M, Cole J, Pollock S et al. (1981) The late appearance of chronic pericardial disease in patients treated by radiotherapy for Hodgkin's disease. Ann Intern Med 94:338–341

Berg NO, Lindgren M (1963) Relation between field size and tolerance of rabbit's brain to roentgen irradiation (200 kV) via a slit-shaped field. Acta Radiol 1:147–168
Bergsagel DE (1971) Total body irradiation for myelomatosis. Br Med J ii:325–327
Bond VP, Fludner TM, Archambeau JO (1965) Mammalian radiation lethality. Academic Press, New York
Brosius FC, Waller BF, Roberts WG (1981) Radiation heart disease: analysis of 16 young (aged 15 to 33 years) necropsy patients who received over 3500 rads to the heart. Am J Med 70:519–530
Byhardt R, Brace K, Ruckdeschel J et al. (1975) Dose and treatment factors in radiation-related pericardial effusion associated with the mantle technique for Hodgkin's disease. Cancer 35:795–802
Carbell SC, Chaffey JT, Rosenthal DS et al. (1979) Results of total body irradiation in the treatment of advanced non-Hodgkin's lymphoma. Cancer 43:994–1000
Catterall M, Evans W (1960) Myocardial injury from therapeutic irradiation. Br Heart J 22:168–174
Choi NC, Kanarek OJ, Kazemi H (1985) Physiologic changes in pulmonary function after thoracic radiotherapy for patients with lung cancer and role of regional pulmonary function studies in predicting postradiotherapy pulmonary function before radiotherapy. Cancer Treat Symp 2:119–130
Cohen KE, Stewart JR, Fajardo LF et al. (1967) Heart disease following irradiation. Medicine 46:281–298
Cohen S, Bharati S, Glass J et al. (1981) Radiotherapy as a cause of complete atrioventricular block in Hodgkin's disease: an electrophysiological-pathological correlation. Arch Intern Med 141:676–679
Constine LS, Donaldson SS, McDougall IR et al. (1984) Thyroid dysfunction after radiotherapy in children with Hodgkin's disease. Cancer 53:878–883
Deeg HJ, Flounoy N, Sullivan K et al. (1984) Cataracts after total body irradiation and marrow transplant: a sparing effect of dose fractionation. Int J Radiat Oncol Biol Phys 10:957–964
Devney RB, Sklar CA, Nesbit ME et al. (1984) Serial thyroid function measurements in children with Hodgkin's disease. J Pediatr 105:223–227
Doll R, Smith PG (1968) The long-term effects of X-irradiation in patients treated for metropathia haemorrhagica. Br J Radiol 41:362–368
Dunsmore LD, LoPonte MA, Dunsmore RA (1986) Radiation-induced coronary artery disease. J Am Coll Cardiol 8:239–244
Evans RG (1983) Radiobiologic considerations in magna-field irradiation. Int J Radiat Biol Phys 9:1907–1912
Fajardo LF (1982) Pathology of radiation injury. Masson Publishing USA Inc, New York
Fajardo LF, Stewart JR (1970) Experimental radiation-induced heart disease. I. Light microscopic studies. Am J Pathol 59:299–316
Fajardo LF, Stewart JR (1973) Pathogenesis of radiation-induced myocardial fibrosis. Lab Invest 29:244–257
Fryer CJ, Fitzpatrick PJ, Rider WD (1978) Radiation pneumonitis: experience following a large single dose of irradiation. Int J Radiat Oncol Biol Phys 4:931–936
Gavin PR, Gillette EL (1982) Radiation response of the canine cardiovascular system. Radiat Res [Suppl] 90:489–500
Glatstein E, Fajardo LF, Brown JM (1977) Radiation injury in the mouse kidney. I. Sequential light microscopic studies. Int J Radiat Oncol Biol Phys 2:933–943
Glatstein E, McHardy-Young S, Brast N et al. (1971) Alterations in serum thyrotropin (TSH) and thyroid function following radiotherapy in patients with malignant lymphoma. J Clin Endocrinol Metab 32:833–841
Hall EJ (1978) Radiobiology for the radiologist. Harper and Row, New York
Heller CG (1967) Effects on the germinal epithelium of radiobiological factors in manned space flight. In: Langham WH (ed) NRC Publication 1487. National Academy of Sciences, National Research Council, Washington DC, pp 124–133
Hindo WA, De Trana F, Lee M-S et al. (1970) Large dose increment irradiation in treatment of cerebral metastases. Cancer 26:138–141
Hoffman WF, Levin VA, Wilson CB (1979) Evaluation of malignant glioma patients during the post-irradiation period. J Neurosurg 50:624–628
Horsey S, White A (1980) Isoeffect curve for radiation myelopathy. Br J Radiol 53:168–169
Hubmann FH (1981) Effect of X-irradiation on the rectum of the rat. Br J Radiol 54:250–254
Ingold JA, Reed GB, Kaplan HS et al. (1965) Radiation hepatitis. Am J Roentgenol 93:200–208
Jaffe JP, Bosch A, Raich PC (1979) Sequential half-body radiotherapy in advanced multiple myeloma. Cancer 43:124–128

Jongejan H, Van der Kogel A, Provoost A et al. (1987) Radiation nephropathy in young and adult rats. Int J Radiat Oncol Biol Phys 13:225–232

Keane TJ, Van Dyk J, Rider WD (1981) Idiopathic interstitial pneumonia following bone marrow transplantation: the relationship with total body irradiation. Int J Radiat Oncol Biol Phys 7:1365–1370

Kemper T, O'Neill R, Caveness W (1977) Effects of single dose supervoltage whole brain radiation in *Macaca mulatta*. J Neuropathol Exp Neurol 36:916–940

Kim TH, Panakon AM, Friedman M (1976) Acute transient radiation hepatitis following whole abdominal irradiation. Clin Radiol 27:449–454

Kim TH, Khan FM, Galvin JM (1980) A report of the work party: comparison of total body irradiation techniques for bone marrow transplantation. Int J Radiat Oncol Biol Phys 6:779–784

Knopse WH, Blom J, Crosby WH (1968) Regeneration of locally irradiated bone marrow. II. Induction or regeneration in permanently aplastic medullary cavities. Blood 31:400–405

Laxton R, Kunkler P (1964) Radiation nephritis. Acta Radiol [Ther] (Stockh) 2:169–178

Leith JT, DeWyngaert K, Glicksman A (1981) Radiation myelopathy in the rat: an interpretation of dose relationships. Int J Radiat Oncol Biol Phys 7:1673–1677

Liebow AA, Warren S, DeCoursey E (1949) Pathology of atomic bomb casualties. Am J Pathol 25:21–26

Lockhart SP, Down JD, Steel GG (1986) The effect of low dose rate and cyclophosphamide on the radiation tolerance of the mouse lung. Int J Radiat Oncol Biol Phys 12:1437–1440

Lushbaugh CG, Casarett GW (1976) The effects of gonadal irradiation in clinical radiation therapy: a review. Cancer 37:1111–1120

Lushbaugh CG, Rider RC (1972) Some cytokinetic and histopathologic considerations of irradiated male and female gonadal tissues. Front Radiat Ther Oncol 6:224–248

Maier JG (1972) Effects of radiations on kidney, bladder and prostate. Front Radiat Ther Oncol 6:196–227

Merriam GR Jr, Focht EF (1957) A clinical study of radiation cataracts and the relationship to dose. Am J Roentgenol 77:759–785

Merriam GR Jr, Szechter A, Focht EF (1972) The effects of ionizing irradiation on the eye. Front Radiat Ther Oncol 6:346–385

Meyers JD, Flournoy N, Thomas ED (1982) Non-bacterial pneumonia after allogeneic marrow transplantation: a review of ten years experience. Rev Infect Dis 4:1119–1132

Nelson DF, Reddy KV, O'Mara RE et al. (1978) Thyroid abnormalities following neck irradiation for Hodgkin's disease. Cancer 42:2553–2562

O'Reilly RJ (1983) Allogeneic bone marrow transplantation: current status and future directions. Blood 62:941–964

Peck WS, McGreer JT, Kretzschman NR et al. (1940) Castration of the female by irradiation. Radiology 34:176–186

Pedrick TJ, Hoppe RT (1986) Recovery of spermatogenesis following pelvic irradiation for Hodgkin's disease. Int J Radiat Oncol Biol Phys 12:117–121

Perez CA, Korba A, Zwnuska F et al. (1978) ^{60}Co moving strip technique in the management of carcinoma of the ovary: analysis of tumor control and morbidity. Int J Radiat Oncol Biol Phys 4:379–388

Perez CA, Stanley K, Rubin P et al. (1980) A prospective randomized study of various irradiation doses and fractionation schedules in the treatment of inoperable non-oat cell carcinoma of the lung. Preliminary report by the Radiation Therapy Oncology Group. Cancer 45:2744–2753

Phillips R, Karnofsky D, Hamelton L et al. (1984) Roentgen therapy of hepatic metastases. Am J Roentgenol 71:826–834

Price RA (1979) Histopathology of CNS leukemia and complications of therapy. Am J Pediatr Hematol Oncol 1:21–30

Reactor safety study (1975) Appendix VI. USNRC, Washington 1400 (NUREG 75/014)

Rider WD (1963) Radiation damage to the brain – a new syndrome. J Can Assoc Radiol 14:67–69

Roswitt B, Malsky S, Reid C (1972) Radiation tolerance of the gastrointestinal tract. Front Radiat Ther Oncol 6:160–181

Rubin P (1973) Regeneration of bone marrow in rabbits following local, fractionated irradiation. Cancer 32:847–852

Rubin P (1977) Radiation toxicity: quantitative radiation pathology for predicting effects. Cancer 39 [Suppl 2]:729–736

Rubin P (1984) The Franz Buschke lecture. Late effects of chemotherapy and radiation therapy: a new hypothesis. Int J Radiat Oncol Biol Phys 10:5–34

Rubin P, Casarett GW (1968) Clinical radiation pathology. Saunders, Philadelphia

Rubin P, Landman S, Mayer E et al. (1973) Bone marrow regeneration and extension after extended field irradiation in Hodgkin's disease. Cancer 32:699–711
Rubin P, Bennett JM, Begg C et al. (1981) The comparison of total body irradiation versus chlorambucil and prednisone for remission induction of active chronic lymphocytic leukemia: an ECOG Study. I. Total body irradiation, response and toxicity. Int J Radiat Oncol Biol Phys 7:1623–1632
Rubin P, Van Houtte P, Constine LS (1982) Radiation sensitivity and organ tolerances in pediatric oncology: a new hypothesis. Front Radiat Ther Oncol 16:62–82
Rubin P, Constine LS, Scarantino CW (1984) The paradoxes in patterns and mechanisms of bone marrow regeneration after irradiation. II. Total body irradiation. Radiother Oncol 2:227–233
Rubin P, Finkelstein JN, Siemann DW et al. (1984) Predictive biochemical assays for late radiation effects. Int J Radiat Oncol Biol Phys 12:469–476
Salazar OM, Van Houtte P, Rubin P (1984) Once a week radiation for locally advanced lung cancer: final report. Cancer 54:719–725
Scarantino CE, Rubin P, Constine LS (1984) The paradoxes in patterns and mechanisms of bone marrow regeneration after irradiation. I. Different volumes and doses. Radiother Oncol 2:215–225
Shalet SM, Beardwell CG, Jacobs HG et al. (1978) Testicular function following irradiation of the human prepubertal testis. Clin Endocrinol 9:483–490
Shalet SM, Horner A, Ahmed JR (1985) Leydig cell damage after testicular irradiation for acute lymphoblastic leukemia. Med Pediatr Oncol 13:65–68
Shank B (1983) Techniques of magna-field irradiation. Int J Radiat Oncol Biol Phys 9:1925–1931
Shapiro E, Kinsella T, Makuch R et al. (1985) Effect of fractionated irradiation on endocrine aspects of testicular function. J Clin Oncol 3:1232–1239
Sheline GE, Waram WM, Smith V (1980) Therapeutic irradiation and brain injury. Int J Radiat Oncol Biol Phys 6:1215–1228
Siemann DW, Hill RP, Penney DP (1982) Early and late pulmonary toxicity in mice evaluated 180 and 420 days following localized lung irradiation. Radiat Res 89:396–407
Sklar CA, Kim TH, Ramsay NKC (1982) Thyroid dysfunction among long-term survivors of bone marrow transplantation. Am J Med 73:688–694
Sklar CA, Kim TH, Williamson JF et al. (1983) Ovarian function after successful bone marrow transplantation in post-menarcheal females. Med Pediatr Oncol 11:361–364
Sklar CA, Kim TH, Ramsay N (1984) Testicular dysfunction following bone marrow transplantation performed during or after puberty. Cancer 53:1498–1501
Speiser B, Rubin P, Casarett G (1973) Aspermia following lower truncal irradiation in Hodgkin's disease. Cancer 32:692–698
Stewart J, Cohen K, Fajardo L et al. (1967) Radiation induced heart disease: a study of twenty-five patients. Radiology 89:302–310
Stewart JR, Fajardo LF (1984) Radiation-induced heart disease: an update. Prog Cardiovasc Dis 27:173–194
Sullivan KM, Deeg HJ, Sanders JE et al. (1984) Late complications after marrow transplantation. Semin Hematol 21:53–63
Tefft M, Mitus A, Das L et al. (1970) Irradiation of the liver in children: review of experience in the acute and chronic phases, and in the intact normal and partially resected. Am J Roentgenol 108:365–385
Thomas ED, Storb R, Clift RA et al. (1975) Bone marrow transplantation. N Engl J Med 292:832–843, 895–902
Thomas ED, Storb R, Buckner CD (1976) Total body irradiation in preparation for bone marrow engraftment. Transplant Proc 8:591–593
Thomas ED, Clift RA, Hersman J et al. (1982) Marrow transplantation for acute non-lymphoblastic leukemia in first remission using fractionated or single-dose irradiation. Int J Radiat Oncol Biol Phys 8:817–821
Trott K-R (1984) Chronic damage after radiation therapy: challenge to radiation biology. Int J Radiat Oncol Biol Phys 10:907–913
Tubiana M, Frindel E, Croizat H (1979) Effects of radiation on bone marrow. Pathol Biol (Paris) 27(6): 326–334
Van der Kogel AJ (1977) Radiation tolerance of the spinal cord: time–dose relationships. Radiology 122:505–509
Wachholz BW, Casarett GW (1970) Radiation hypertension and nephrosclerosis. Radiat Res 41:39–56
Wara WM, Phillips TL, Margolis LW et al. (1973) Radiation pneumonitis: a new approach to the derivation of time–dose factors. Cancer 32:547–552

White DC (1975) An atlas of radiation histopathology. ERDA, TID-26676, Oak Ridge, Tennessee
Willett C, Tepper J, Orlow E et al. (1986) Renal complications secondary to radiation treatment of upper abdominal malignancies. Int J Radiat Oncol Biol Phys 12:1601–1604
Woods W, Dehner L, Nesbit M (1980) Fatal veno-occlusive disease of the liver following high-dose chemotherapy, irradiation and bone marrow transplantation. Am J Med 68:285–290

11 Total Body Irradiation: Clinical Aspects

A. Barrett

Introduction

Total body irradiation (TBI) is being used increasingly as consolidation treatment in the management of leukaemia, lymphoma and various childhood tumours with the aim of sterilizing any residual malignant cells or micrometastases.

Systemic radiotherapy as an adjunct to chemotherapy offers several possible benefits. There are no sanctuary sites for TBI; some neoplastic cells are very radiosensitive, and resistance to radiation appears to develop less readily than to drugs. Cross-resistance between chemotherapy and radiotherapy does not seem to be common and although plateau effects may be seen with chemotherapy there is a linear dose–response curve for clonogenic cell kill with radiation.

Most clinical experience with total body irradiation has been in the treatment of leukaemia, a disease where the target cell is assumed to have the same high degree of radiosensitivity as its normal counterpart. For other tumours the target cell will be relatively more resistant than haemopoietic cells, and bone marrow transfusion is then given to rescue the patient from the total body dose-limiting effects of bone marrow ablation.

Radiation Dose

It has been considered by many that homogeneity of dose distribution throughout the body is desirable (by analogy with conventional small-volume treatments). The use of bolus material or compensators at sites such as the ankles and neck has therefore been advocated where doses would otherwise be up to 10%–20% higher than at the midplane of the abdomen (Kim et al. 1980). Since damage in these areas is unlikely at the doses prescribed, this attempt to achieve homogeneity seems superfluous. The aim of TBI should rather be the delivery of as high a dose as possible to any probable tumour site, perhaps with selective shielding of critical organs. These, in order of sensitivity, might be the lungs, kidneys, liver and brain. The risk of gut damage is difficult to separate from bone marrow injury but would probably also be dose-limiting at levels near to those that would cause damage to the lung.

Accurate shielding of these organs under usual TBI conditions is rarely feasible and in practice the need to restrict lung damage determines the dose given. It is not always appreciated that mediastinal radiation (as for example if a boost of total lymph node irradiation is given) will also increase the risk of pneumonitis. Because of the difficulty of accuracy with long single-dose treatments and the possibility of leukaemic infiltration of ribs and lung tissue, we have preferred to limit our total TBI dose to that of lung tolerance and not to use any shielding at all.

Accurate dosimetry is of paramount importance since an increase of only 10% may dramatically affect the incidence of lung problems. In vivo measurement with lithium fluoride thermoluminescent dosimetry seems the best method at present for measuring dose distribution, though diodes may also be satisfactory (Lui et al. 1983). In some centres in the United States, methods of dose calculation are preferred (Khan et al. 1980; Galvin et al. 1983). The data on which such calculations are based are not always derived under TBI treatment conditions (i.e. at extended focus to skin distance using an infinite phantom, etc.) and unless individual measurements are made for each patient, inaccuracy may occur.

A maximum variation of dose of ±5% over the body (apart from neck and ankles) can be achieved with an accuracy of measurement of ±2.5% (Lam et al. 1979).

For single-fraction TBI, dose rate has been shown to be a critical factor. Animal studies have demonstrated reduced acute toxicity when dose rates below 4 cGy/min are used (Depledge et al. 1982), although no further benefit can be obtained by reduction to 0.4 cGy/min (Kolb et al. 1987). In man the risk of interstitial pneumonitis after TBI is correlated with dose rate, again with a lower risk at dose rates below 4 cGy/min (Barrett 1982). Tolerance doses have been clearly defined for two dose schedules. At the Royal Marsden Hospital, Surrey, an escalating dose study using dose rates of 4 cGy/min showed no difference in toxicity between 9.5 and 10.5 Gy but a steep rise in the incidence of death, both from all causes and from lung problems in particular, when doses of 11.5 Gy were used (Barrett et al. 1987). A similar attempt at the Royal Free Hospital, London, to increase dose from 7.5 to 8 Gy at a dose rate of 25 cGy/min again caused a steep rise in complication rate (G. Prentice, personal communication). Increasing dose brought an increase in actuarial leukaemia-free survival but the benefit of this at 11.5 Gy was offset by the increased toxicity so that the optimal dose level in practice was 10.5 Gy.

No single-treatment doses higher than this have been reported and most centres use lung shielding at 8–9 Gy. The rate of pneumonitis after 7.5 Gy at high dose rates is similar. Unfortunately no comparisons of leukaemic relapse rates have been made so it is difficult to know whether there is any advantage to either regime. Total dose seems to be important (Kolb et al. 1987) and in theory dose rate should be relatively insignificant for leukaemic cell killing, but this remains to be proven.

Because of logistic problems in the organization of long single treatments, most centres have adopted a fractionated schedule. This was proposed by the Seattle group after a randomized study which showed improvement in survival after six fractions of 2 Gy given over 3 days compared with 10 Gy at a dose rate of 0.08 Gy/min (Thomas et al. 1982). The widespread use of this schedule has reduced the rate of interstitial pneumonitis from the high rates seen after single-fraction high-dose-rate treatments, although no improvement over single-fraction low-dose-rate treatments has been demonstrated. No conclusions can be drawn about rates of leukaemia control but no striking differences between schedules have been noted. Several other fractionation schemes have now come into use which are unlikely to offer the same sparing effects. In particular those with very large doses per fractionation (5 Gy) and those with only a few hours between fractions seem unsatisfactory on theoretical grounds.

The recent introduction of T cell depletion of bone marrow in an attempt to prevent graft versus host disease has been associated with a high rate of graft rejection (Maraninchi et al. 1987). It was postulated that this effect was due to failure to eliminate a radioresistant cell population (perhaps NK cells) which would have been removed previously by donor T cells. An increase in radiation dose was recommended to overcome this problem (Slavin et al. 1985). Some early success was claimed but evidence now is conflicting, and there are objections to such claims. In many cases the proposed increase in dose of fractionated treatment may not in fact have represented a biologically more effective dose than the single-fraction regimes used previously, and the cell responsible for rejection has not yet been characterized. T cell depletion has led to a higher rate of leukaemic relapse and current effort is directed towards achieving partial T cell depletion, which will minimize graft versus host disease and maximize control of leukaemia.

Toxicity

The acute radiation effects of nausea and vomiting have been described previously (Westbrook et al. 1987). There is an apparent threshold at about 2 Gy. Young age at treatment reduces the incidence, which is increased by vestibular disturbance and sensory input via the vagus. Vomiting can be prevented completely (by adequate sedation) in up to 50% of patients receiving 10 Gy single-fraction TBI. Diarrhoea at 3–4 days is usually not severe but is worsened when certain chemotherapeutic agents such as melphalan and methotrexate are used in combination with TBI. Hypotension and pyrexia can be abolished by administration of steroids, and parotitis, associated with a rise in amylase levels for 36–48 hours, is not usually severe.

In the period up to 100 days after transplantation pneumonitis is the most significant clinical problem (Neiman et al. 1978). It is multifactorial in origin, the most important predisposing factors being radiation, graft versus host disease and infection, particularly with cytomegalovirus or *Pneumocystis*. The mortality rate of the established process is high.

Somnolence is also seen at 6–8 weeks after TBI and, as with the syndrome seen after prophylactic cranial irradiation, usually resolves spontaneously within a week.

Late effects (Barrett 1987) may include cataract, which occurs in from 20% to 80% of patients (depending on the radiation technique used) within 3–4 years of transplantation. Surgical removal may be undertaken in most patients with good effect.

Effects on hormonal function are the earliest of the late effects to become obvious. Ovarian failure occurs in nearly all females with low circulating oestradiol and raised follicle stimulating hormone (FSH) levels. Azoospermia occurs in men and recovery has not yet been reported. FSH levels are raised but testosterone levels are normal. Puberty is delayed in nearly all girls and some boys. Growth is impaired by two processes. A direct effect of radiation on the epiphyses produces failure of growth and leads to a reduction in sitting height relative to total height, and in patients who have had previous prophylactic cranial irradiation growth hormone impairment may also contribute to growth failure. Overt hypothyroidism is uncommon but compensated hypothyroidism with normal T_3/T_4 levels and raised thyroid stimulating hormone is seen in some patients.

Second tumours are so far an uncommon problem, with an incidence of about 1%–2%. Relapse or second leukaemias are most frequently seen. Lymphoproliferative disorders, including non-Hodgkin's lymphomas, occur at 4–5 years after bone marrow transplantation and solid tumours have been observed in small numbers. These include skin and brain tumours and intestinal adenocarcinoma (Buckner 1987). Data from animal studies (Kolb et al. 1987) and experience after the bomb explosions in Hiroshima and Nagasaki (Hiroshima and Nagasaki 1981) suggest that the incidence may be expected to rise with time.

No other significant late effects have been reported yet, though there is a possibility that vascular changes developing 10–20 years after TBI may produce clinical damage in organs such as the kidney, brain and liver which have received near-tolerance doses. Careful follow-up of all patients treated with TBI is essential for monitoring late effects and, where feasible, for reducing their incidence.

References

Barrett A (1982) Total body irradiation (TBI) before bone marrow transplantation in leukaemia: a co-operative study from the European Group for Bone Marrow Transplantation. Br J Radiol 55:562–567

Barrett A (1987) Late effects of total body irradiation. Radiat Oncol (in press)

Barrett A, Dobbs J, Milan S, Powles R, Westbrooke K (1987) Death after bone marrow transplantation – relationship to total body irradiation. (Submitted)

Buckner CD (1987) Presentation, European Bone Marrow Transplantation Group XIII meeting. Interlaken, March 1987

Depledge MH, Barrett A (1982) Dose rate dependence of lung damage after total body irradiation in mice. Int J Radiat Oncol Biol Phys 41:325–334

Galvin JM (1983) Calculation and prescription of dose for total body irradiation. Int J Radiat Oncol Biol Phys 9:1919–1924

Hiroshima and Nagasaki (1981) The committee for the compilation of materials on damage caused by the atomic bombs in Hiroshima and Nagasaki. Iwanami Shoten Publishers, Tokyo

Khan FM, Williamson JF, Sewchand W, Kim TH (1980) Basic data for dosage calculation and compensation. Int J Radiat Oncol Biol Phys 6:745–751

Kim TH, Khan Faiz M, Galvin JM (1980) A report of the work party: comparison of total body irradiation techniques for bone marrow transplantation. Int J Radiat Oncol Biol Phys 6:779–784

Kolb HJ et al. (1987) Dose rate and fractionation studies of TBI in dogs – short and long term effects. European Bone Marrow Transplantation Group XIII meeting, Interlaken, March 1987

Lam WC, Lindskoung BA, Order SE, Grant DG (1979) The dosimetry of ^{60}Co total body irradiation. Int J Radiat Oncol Biol Phys 5:905–911

Lui JC, Bacza ET, Findley DO, Forell BW (1983) Dosimetry of single fraction high dose total body irradiation as measured by thermoluminescent dosimeters. Int J Radiat Oncol Biol Phys 9:1407–1408

Maraninchi D, Gluckman E, Blaise D, et al. (1987) Impact of T cell depletion on results of HLA matched bone marrow transplantation in standard risk malignancies: a randomised study of the GEGMO (French Cooperative Group). European Bone Marrow Transplantation Group XIII meeting, Interlaken, March 1987

Neiman P, Wasserman PB, Wentworth BB, et al. (1978) Interstitial pneumonia and cytomegalovirus infection as complications of human marrow transplantation. Transplantation 15:5

Slavin S, Or R, Weshler Z, Fuks Z, Morecki S (1985) Radiation for allogeneic bone marrow transplantation in animals and man. Surv Immunol Res 4(3):238–252

Thomas ED, Clift RA, Hersman J et al. (1982) Marrow transplantation for acute non-lymphoblastic leukaemia in first remission using fractionated or single-dose irradiation. Int J Radiat Oncol Biol Phys 8:817–821

Westbrook C, Glaholm J, Barrett A (1987) Vomiting associated with whole body irradiation. Clin Radiol (in press)

12 Particle Therapy: Physics and Biology

D. K. Bewley

Introduction

The term particle therapy is used to describe treatment with the more exotic types of radiation, excluding X-rays, gamma-rays and electrons. Their value for therapy depends as much on their physical properties as on their radiobiology. These aspects are so interwoven that one cannot be discussed without the others (Raju 1980; Fowler 1981).

The particles to be discussed are listed in Table 12.1 along with an outline of their characteristics which are relevant to radiotherapeutic use. Neutrons are uncharged and are attenuated in tissue more or less exponentially, like X- and gamma-rays. The charged particles have a definite range in tissue and travel in nearly straight lines. As a result the absorbed dose is almost zero beyond the range and the beam has a very sharp penumbra. Negative pi-mesons (pions) are light enough to show some of the characteristics of electrons, i.e. considerable scattering and an imprecise range. But they have the special property that when reduced to almost zero velocity each pion causes a nuclear disintegration with emission of charged particles and neutrons.

Table 12.1 also gives the energies and currents of the charged particles for treatment to a depth of 20 cm. The energies are high, hundreds or thousands of megaelectron-volts, and the currents very small. Large and expensive accelerators are needed to achieve these energies. Quite high energies are needed even for superficial sites; for example the treatment of choroidal melanomas needs at least 60-MeV protons. Neutrons and pions are not accelerated directly but are produced at a target, like X-rays, and so require much larger currents from the accelerator. For a good therapeutic beam of fast neutrons one needs about 20 μA of 50-MeV protons or deuterons. Pions are more difficult to produce; even when they are collected with maximum efficiency, as at the Swiss Institute for Nuclear Research (SIN), it is necessary to accelerate 20 μA of protons at 590 MeV onto the target.

Table 12.1. Characteristics of the particles used in particle therapy

Particle:[a]	π	n	p(H)	He	C	Ne	Ar
Mass	0.15	1	1	4	12	20	40
Charge	−1	0	1	2	6	10	18
MeV/nucleon[b]	—	—	175	175	330	470	630
GeV/particle[b]	0.082	—	0.17	0.7	4.0	9.4	25
Min. current (pA)[c]	~10	—	200	100	50	35	25
Entrance LET[d] (keV/μm)	~2	1–1000	0.5	2	20	35	80
Max. LET at depth (keV/μm)	~500	~1000	100	250	700	1000	>1000

[a] π, negative pi-meson; n, neutron; p(H), proton; He, helium; C, carbon; Ne, neon; Ar, argon.
[b] Energy for a range of 20 cm in water. 1 GeV = 1000 MeV.
[c] Minimum current of fully stripped ions to give 1 Gy/min over a volume of 10 cm × 10 cm × 20 cm deep.
[d] LET, linear energy transfer.

Physical Aspects of Particle Beams

Charged ions deposit energy at a steadily increasing rate until very close to the end of the range (the Bragg peak). To obtain a uniform dose over a useful thickness of tissue, approximating the thickness of the tumour, the energy of the beam has to be modulated so that the position of the peak spans the depths required. The principle is explained in Fig. 12.1. Dose distributions based on this principle are superior to anything that can be obtained with X- or gamma-rays, neutrons or electrons. Therapy with charged ions is particularly useful for treating disease close to vital normal structures such as the spinal cord (Austin-Seymour et al. 1985). With increasing mass of the ion, scattering is reduced. This

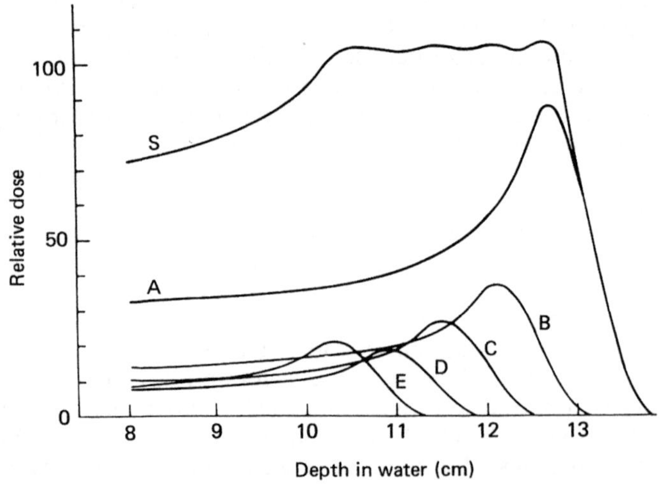

Fig. 12.1. Depth–dose distribution from protons showing the extended peak produced by energy modulation. Protons at the full energy (*A*) are added to contributions at lower energies (*B* to *E*) to give the sum *S*. (After Raju 1980.)

is useful when very small fields are used as in stereotactic surgery. On the other hand, heavier ions suffer more nuclear interactions during their passage through tissue. This decreases the ratio of peak dose to entry dose and produces long-range fragments which cause some dose to be deposited beyond the range of the parent ions. With all these ions, shaped filters (tissue compensators) can be used to adjust the range of the particles in different parts of the field, giving complex three-dimensional high-dose volumes.

Negative pions give a less sharp fall-off beyond the peak, partly because fast neutrons are produced after capture of the stopping pions and partly because pion beams are always contaminated with other species such as muons and electrons. Fast neutrons, being uncharged, are attenuated more or less exponentially and confer no advantage in terms of dose distributions. They differ from X- and gamma-rays in that deposition of energy depends largely on the hydrogen content of the tissue. Thus fast neutrons deposit less energy in bone and more in fat – about 70% and 115% respectively of the energy deposited in muscle. This is useful in avoiding bone necrosis when tumour lies close to bone, but is disadvantageous in that it leads to more subcutaneous fibrosis (Catterall and Bewley 1979).

Radiobiology

LET Distributions

Except for protons, these particles constitute radiations with a substantially different quality from X- and gamma-rays. Quality is usually described in terms of the linear energy transfer (LET) of the charged-particle tracks. The LET of the charged ions rises steadily as they pass into tissue, reaching a maximum in the Bragg peak, so that the LET is greater at a depth (that is in the tumour) than at the entry point of the beam. Table 12.1 gives values for the LET of the various particles at the entrance and the maximum attained in the Bragg peak. The top of the peak is so narrow that the maximum values in Table 12.1 occur only over a depth of a few millimetres. When the Bragg peak is spread out as in Fig. 12.1 the maximum LET only occurs at its distal end. Pions behave similarly except for an additional high LET component in the peak which arises from the products of nuclear disintegration. In the case of neutrons, radiation quality does not change appreciably with depth but does depend on neutron energy. Neutrons produce a variety of charged secondary particles of low energy: protons, helium ions (α particles), heavy recoils of carbon and oxygen nuclei and products of nuclear reactions. The LET along the tracks of these particles extends over a wide range, mostly between 5 and 500 keV/μm. Fig. 12.2 shows LET spectra for two neutron energies and for ^{60}Co gamma-rays.

Fig. 12.3 indicates how the relative biological effectiveness (RBE) and the oxygen enhancement ratio (OER) vary with LET. Over the range of radiation quality included in beams of X- and gamma-rays there is little change in these parameters. The particle beams, on the other hand, span a region where both vary rapidly. In the case of the charged ions over the range of atomic numbers from helium to beyond neon, the RBE at a depth is greater than at the surface.

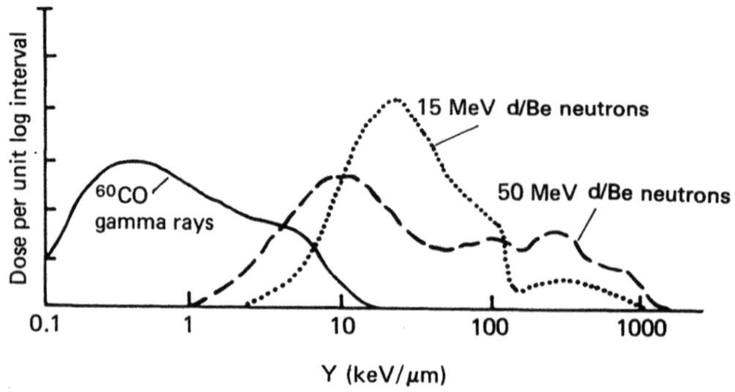

Fig. 12.2. Spectra of lineal energy Y (similar to LET) for two neutron beams and ^{60}Co gamma rays. (After Ito 1980.)

The effective dose (dose × RBE) at a depth relative to the surface is therefore even greater than is given by the physics alone – an advantage for therapy with those ions. The same is true for pions. However, this is not true for hydrogen ions or ions of argon and heavier. With argon ions the RBE is already nearly at its maximum when they enter the patient and a further increase in LET causes a fall in RBE. With hydrogen ions the LET exceeds 30 keV/μm only very near the end of the track; when the Bragg peak is broadened by velocity modulation the increased RBE in the peak is almost completely lost.

Radiobiological Effects at Raised LET

In general the damage caused by radiation becomes less susceptible to modifying factors as the LET rises. The radioresistance caused by anoxia is an example and is indicated in Fig. 12.3. Other examples are the variation in radiosensitivity of

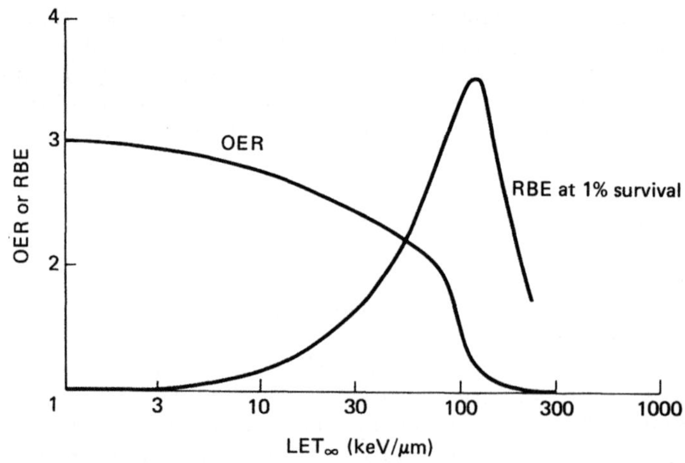

Fig. 12.3. Schematic diagram to show the variation in RBE and OER with LET. RBE also varies with end-point and with dose per fraction.

cells as they progress round the cell cycle and the capacity for repair of sublethal and potentially lethal damage. Beyond an LET of 200 keV/μm the RBE falls because the tracks are so densely ionizing that more energy is deposited when a track intersects a sensitive site than is needed to inactivate the cell.

High LET is only useful if the RBE of the radiation is greater for the tumour than for the limiting normal tissues. Also there is little point in using a new and expensive form of radiation to treat disease which responds well to conventional radiation. The problem is to identify the factors which cause certain tumours to be radioresistant and to discover whether radiation of higher LET would have a greater RBE for destroying them than for damaging relevant normal tissues.

The first of these factors to be discussed is hypoxia, the original rationale for introducing high LET radiation. Treatment under high-pressure oxygen has been found to increase the success rate of radiotherapy in certain sites (Dische 1979; Henk 1986). A retrospective analysis of anaemic patients suffering from stage III cancer of the cervix showed a particularly striking increase in effectiveness, persisting local control rising from 20% to 90% (Dische et al. 1983). This is the first prognostic indicator to be found which identifies a small subgroup of patients who should benefit by treatment with high LET radiation. Hypoxia may also be important in very large tumours and in tumours which have recurred after previous radiation treatment.

Variations in radiosensitivity round the cell cycle have been studied by many workers and the fact that these variations are reduced with high LET radiation is well known. It is doubtful, however, whether this has much therapeutic significance, partly because tumour cells are reassorted between fractions and partly because the cells of normal tissues also show variations in radiosensitivity.

Repair of sublethal and potentially lethal damage is likely to be a more important factor. It is almost a universal finding that repair is reduced after irradiation at increased LET. This is the main reason why RBE rises as the dose per fraction is reduced. However, high LET radiation will be beneficial only when the tumour cells show a larger capacity for repair after X-rays than do the cells of limiting normal tissues. Favourable examples are probably melanoma (Trott et al. 1981) and other radioresistant tumours, whereas unfavourable normal tissues include kidney and spinal cord.

Battermann et al. (1981) found that the RBE of fast neutrons for control of metastases in the lung depended on the growth rate of the tumour, slowly growing tumours showing a higher RBE. This is probably an example of increased repair of damage after irradiation by X-rays in cells not called upon to divide frequently. This may be the reason for the effectiveness of fast neutrons in controlling slowly growing tumours such as those in the salivary glands (Catterall and Bewley 1979). It is also likely to be the reason for the high RBE which has often been noted for late morbidity after neutron therapy in tissues where cells normally have a low division rate.

There is also a rationale for treating tumours in shorter overall times with high LET radiation. Short treatment schedules may not allow adequate time for reoxygenation – important for X-rays but less so for high LET radiation. There is considerable evidence that tumours can often repopulate to a surprising degree during conventional treatment regimes, particularly during weekends and other gaps in treatment (Maciewski et al. 1983; Denekamp 1986). Short treatments are economical, offsetting the disadvantage of the cost of heavy-particle accelerators.

Conclusion

Particle therapies have a number of advantages and disadvantages as alternatives to conventional radiation. These are summarized in Table 12.2. The factors may be physical or radiobiological or a combination of the two. The only advantage of protons lies in the depth–dose distributions which can be obtained. Heavy ions and pions have this advantage and also give beams with higher LET at a depth than at the surface. Neutrons have a physical advantage when bone is closely involved but any other advantage must be radiobiological. High LET radiation can be advantageous for treating tumours containing radioresistant cells, whether this is due to hypoxia or other reasons. Disadvantages include the cost and complexity of equipment and late morbidity arising in certain tissues after irradiation at high LET.

Table 12.2. Advantages and disadvantages of particle therapy

Rationale	Probable examples
For	
Dose distributions of charged ions and pions: limited range and sharp beam edges	Ocular melanomas, brain surgery, tumours very close to sensitive vital tissues such as spinal cord
Reduced neutron kerma in bone	Nasal and oral cancer (neutrons)
Hypoxic tumour cells	Anaemia, very advanced tumours, sarcomas, re-irradiation
Slowly growing tumours	Salivary glands, prostate
Intrinsically radioresistant tumours	Melanoma, sarcoma
Against	
Expense and complexity	Small departments and poor countries
Tumours sensitive to X- and gamma-rays	Testis, lymphoma
Limiting normal tissue spared by fractionation with X- and gamma-rays	Spinal cord, kidney, larynx, lung
Increased neutron kerma in fat	Subcutaneous fibrosis after low-energy neutron therapy

References

Austin-Seymour M, Munzenrider JE, Goiten M et al. (1985) Progress in low-LET heavy particle therapy: intracranial and paracranial tumors and uveal melanomas. Radiat Res 104:S219–S226

Battermann JJ, Breur K, Hart GAM et al. (1981) Observations on pulmonary metastases in patients after single doses and multiple fractions of fast neutrons and cobalt-60 gamma rays. Eur J Cancer 17:539–548

Catterall M, Bewley DK (1979) Fast neutrons in the treatment of cancer. Academic Press, London

Denekamp J (1986) Cell kinetics and radiation biology. Int J Radiat Biol 49:357–380

Dische S (1979) Hyperbaric oxygen: the Medical Research Council trials and their clinical significance. Br J Radiol 51:888–894

Dische S, Anderson PJ, Sealy R et al. (1983) Carcinoma of the cervix – anaemia, radiotherapy and hyperbaric oxygen. Br J Radiol 56:251–255

Fowler JF (1981) Nuclear particles in cancer treatment. Adam Hilger, Bristol
Henk JM (1986) Late results of a trial of hyperbaric oxygen and radiotherapy in head and neck cancer: a rationale for hypoxic cell sensitizers. Int J Radiat Oncol Biol Phys 12:1339–1341
Ito A (1980) Microdosimetry of high LET therapeutic beams. In: Booz J, Ebert HG, Hartfiel HD (eds) Proceedings of 7th symposium on microdosimetry. Harwood Academic Publishers, EEC, pp 1191–1200
Maciewski B, Preuss-Bayer G, Trott KR (1983) The influence of the number of fractions and of overall treatment time on local control and late complication rate in squamous cell carcinoma of the larynx. Int J Radiat Oncol Biol Phys 9:321–328
Raju MR (1980) Heavy particle therapy. Academic Press, New York
Trott KR, Von Lieven H, Kummermehr J et al. (1981) The radiosensitivity of malignant melanomas. 1. Experimental studies. Int J Radiat Oncol Biol Phys 7:9–13

13 Clinical Aspects of Particle Therapy

H. M. Warenius and R. D. Errington

Introduction

Failure to control local disease is still a major cause of death in a large number of common cancers (Table 13.1). In addition to contributing to mortality itself uncontrolled local disease is often the cause of considerable distress to the patient and thus results in poor quality of life. There is therefore still a great need to improve tumour control at the primary site.

Table 13.1. Estimate of the total number of local failures in the US cancer population

Tumour site	Annual deaths 1982[a]	Estimated no. of patients with local failure as major cause of death
Head and neck	9150	3660
Oesophagus	8300	4370
Breast	37300	5260
Cervix uterus	7100	4260
Corpus uterus	3000	1710
Ovary	11400	9600
Prostate	23300	14120
Bladder	10600	5300
Brain, CNS	10400	9710
Skin	6900	4830
Lung	111000	10090
Lymphoma	21200	2560
Total	259550	75470

[a] American Cancer Society: *Cancer Facts and Figures, 1982* (from Seydel 1984).

Since the first clinical studies by R. S. Stone in the 1940s, particle therapy has been proposed and investigated as a possible means of achieving better local control of tumours (Stone and Larkin 1942; Stone 1948). Both the studies of the past and proposed future studies, however, should be seen in context.

Whilst continuing progress has been made in heavy particle irradiation by its protagonists, other workers have attempted to exploit different methods of improving the local control of cancer. Hypoxic cell sensitizers provide a potential method of overcoming clinically the hypoxic component of many cancers, long thought to play an important part in their radioresistance (see Chaps. 14 and 15). In addition, photon therapy when used alone with conventional fractionation has been shown to achieve extremely good results in clinical situations such as colonic adenocarcinoma (Overgaard et al. 1984) previously thought to be less amenable to conventional photon radiotherapy. An improved understanding of fractionation (Thames et al. 1983) is leading to further clinical studies using accelerated fractionation and hyperfractionation treatment schedules.

What is an Appropriate Control Arm in Studies of Particle Therapy?

The importance of the alternative therapeutic strategies described above lies in deciding which of them provides an appropriate control arm in studies involving particle therapy. If conventional photon therapy is used as a control arm at the outset of a study lasting 5–7 years, how can that study best be evaluated on completion when improvements in the conventional photon treatment alone may have occurred? This leads to specific questions such as whether particle therapy using neutrons should be compared with the next generation of hypoxic cell sensitizers or whether neutrons for head and neck cancer should be compared with photon therapy alone, or with photon therapy plus chemotherapy. Also, should neutron therapy in some sites be compared with conventional fractionation, hyperfractionation or accelerated fractionation? One approach to this problem is the design of three-limb studies in which two experimental modalities are compared with each other and the conventional treatment of the time. This, however, may require prohibitively large numbers of patients in most instances.

Evaluation of Particle Therapy

When evaluating particle therapy itself, it should also be borne in mind that considerable changes in techniques are still occurring. Thus the global use of the term "fast neutron" can obscure the fact that different facilities have neutrons of different energy, that these facilities may be hospital-based or non-hospital-based, and that they may produce fixed or isocentric beams. Such considerations can make considerable difference to the clinical outcome (Catterall 1982a,b) and in part may both reflect the varied results of the past and give hope for more consistent successful results in the future. Similarly there have been developments in proton therapy in terms of physical distribution and dosimetry allowing the present treatment of ocular tumours (Goiten and Miller 1983) with a great degree of precision using proton beam energies of 60–70 MeV. Protons are being accelerated to 175–200 MeV to produce suitable beams for treating intra-abdominal and intra-thoracic sites, and this work at Harvard has been reviewed (Munzenrider 1985).

Clinical Reality

Another important aspect of the context in which particle irradiation should be seen is that of clinical reality. Each patient is an individual and tumour behaviour varies from patient to patient. Tumour cell populations are usually heterogeneous with differing phenotypes within the whole population of cells. Such variability in the clinical context may result in subgroups of cells which are potentially responsive to the modality under investigation but are too small to be clinically detected unless very large numbers of patients are treated. This practical statistical problem may explain the conflict between the anecdotal observations of striking responses where the majority of clinicians would agree success was extremely unlikely and the failure to detect such small subgroups in the overall results of randomized controlled trials. Better methods of understanding those characteristics of individual tumours are needed which will enable prediction of the appropriate treatment modality for each subgroup. An important advance along this line is recent work (McNally and Wilson 1986) involving bromodeoxyuridine incorporation and flow cytometry to measure the kinetics of tumour and normal tissue cells which can be applied to estimate the potential doubling time of human tumours (see also Chap. 7). These values may then be correlated with the clinical results observed from the different radiation treatment modalities used. It must not be overlooked that what is sought is an improvement for the average patient presenting at the average oncology centre rather than what may be artificial differences between highly selected groups of patients entered into clinical studies.

Finally, there is the problem of the failure to get general agreement with regard to time, dose and fractionation between many radiotherapy centres. This makes it difficult to select an appropriate photon control arm. It may in fact be easier to decide on an optimal neutron treatment than to decide on the optimal photon treatment to be given as the conventional treatment control arm in such studies.

Charged Particle Therapy

The rationale for the use of charged particle and fast neutron beams has been reviewed by Suit and Goiten (1980). The possibility of improved results with

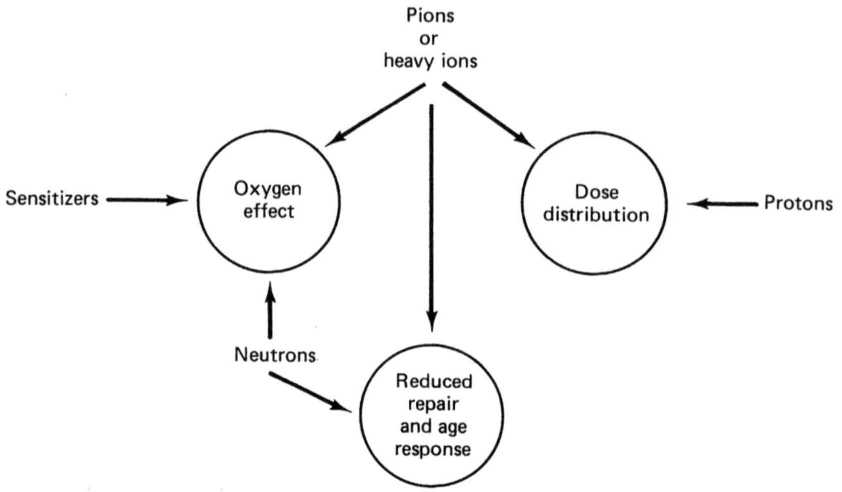

Fig. 13.1. The problems addressed by the various new modalities used or proposed for clinical radiotherapy. The use of electron-affinic radiosensitizers is based on the need to reduce the oxygen effect; protons improve the dose distribution; the use of neutrons is based on a reduced oxygen effect, a reduction of sublethal damage repair and reduced age-response function; pions and heavy ions address all three problems. (From Hall 1981.)

particle radiations is based on biological and physical criteria. Protons and helium ions are low LET (linear energy transfer) and intermediate LET radiations respectively, as are megavoltage photons. Neutrons, negative pi-mesons (pions) and heavy ions (carbon, neon and argon) are high LET radiations. The last group are more efficient biologically than low LET radiations due to a decrease in the repair capacity of the cell and its age-response function. The oxygen enhancement ratio (OER) of higher LET radiation is also reduced, so enhancing the degree of hypoxic cell killing.

Compared with megavoltage X-rays, neutrons have potential biological advantages (Field and Fowler 1984). This is not the case with protons and helium ions but their advantage is that they produce better physical dose distributions. Pions and ions heavier than helium (carbon, neon and argon) have both biological and dose distribution advantages. These new modalities have been reviewed by Hall (1981), and Fig. 13.1 illustrates how they relate to the problems they address.

Pion Therapy

Pions have similar properties to beams of charged particles and have a mass intermediate between that of the electron and the proton. They have a finite range but when brought to rest deposit an additional dose of densely ionizing radiation around the Bragg peak. They therefore offer the possibility of concentrating energy, some of it high LET, with a reduced OER and increased biological effectiveness within the tumour volume. The great disadvantage of pions is that they require an elaborate and expensive accelerator for their production.

Long-term clinical results from Los Alamos (LAMPF) reported by Von Essen et al. (1986) showed some favourable (squamous carcinoma and adenocarcinoma) and some disappointing (pancreatic tumours and gliomas) responses. In all categories a total of 227 patients were treated between 1974 and 1981; there were insufficient numbers in each group to refine optimum dose-fractionation schedules.

At the Vancouver (TRIUMF) pion facility patients with grade 3 and 4 astrocytoma receiving 33 Gy of pions to a limited volume had a marginal survival advantage compared with patients treated with conventional photons. Unacceptable morbidity was seen in rectal carcinoma patients but some early encouraging results were seen in prostatic carcinoma.

At the Swiss Institute for Nuclear Research (SIN) pion therapy continues (Greiner et al. 1986). No benefit has been seen in using pions to treat high-grade malignant gliomas. A dose of 33 Gy in 20 fractions over 5 weeks controlled 50% of advanced bladder tumours. Keeping the target volume as small as possible was critical if severe morbidity was to be avoided. The best responses were seen in soft tissue sarcomas.

Charged Ions

The heavier ions have almost straight tracks and a finite range, depositing energy at an increasing rate as they are slowed down in tissue. The maximum dose occurs at the end of the track (Bragg peak) and by modulation of the energy of incident ions this peak can be spread out to selectively encompass a tumour. Almost ideal dose distributions can therefore be obtained. This is in sharp contrast to the situation with X-rays and neutrons where exponential attenuation greatly restricts the ability to irradiate a tumour selectively.

A great limiting factor in charged ion therapy is again the expensive high-energy accelerators needed. For example protons need to be accelerated to 175 MeV to have a range of 20 cm in tissue. Heavier ions have to be accelerated to greater energies though they do have an advantage over protons: their tracks are straighter and ionization density along them is greater, resulting in greater biological effectiveness in the Bragg peak region. This enhances tumour effects and spares normal tissues nearer the surface which receive sparsely ionizing radiation.

Treatment with charged ions is not widely available though the limited clinical experience to date is encouraging (Castro et al. 1985). Bewley (1987) has outlined a proposal that would lead to a proper trial of this technique. This is the EULIMA (European light-ion medical accelerator) project. In a purpose-built hospital-based accelerator ions in the hydrogen to neon range would be accelerated to energies high enough to give up to 20-cm penetration in tissue.

Proton Therapy

The main advantage of protons in clinical radiotherapy is the improved dose distribution possible because of the physical characteristics of the beam. A greater dose to the target volume can be given with the proton beam than with comparable photon techniques whilst giving the same or an even smaller dose to adjacent normal tissues.

At Harvard fractionated precision high-dose proton therapy has been carried out since 1973 and 846 patients have been treated (Munzenrider et al. 1985). Normal tissue and tumour responses were consistent with an RBE of 1.1 for the proton beam. From the results of these non-randomized studies it has been suggested that proton beam therapy is the treatment of choice for patients with uveal melanomas, chordomas and chondrosarcomas involving the skull base and cervical spine. In patients with prostatic carcinoma, head and neck malignancies, ano-rectal cancers and retroperitoneal tumours the improved dose distributions possible with protons have allowed greater doses to be delivered than would be given conventionally with photons. Proton doses were 10%–20% greater than would normally be given as radical photon therapy and local control rates have been good.

Between 1977 and 1984, 524 patients with uveal melanoma were treated with protons at Harvard. Overall 89% of 332 tumours decreased in size after treatment and no regrowth in the treated area was observed in 476 of 484 eyes (98%) observed between 6 and 108 months after treatment. The majority of the total group of patients (60%) had tumours that were large (maximum diameter greater than 15 mm and/or height greater than 5 mm). Only 6.3% of patients have developed metastases and most of these were in the patients presenting with large tumours. Useful vision was preserved after treatment in the majority of patients. Approximately two-thirds of treated eyes had vision the same or better than that present initially, with a lesser degree of visual preservation in eyes with larger tumours.

In view of these impressive results from non-randomized studies a study is to be started using the Clatterbridge cyclotron, Merseyside, to treat uveal melanomas. This cyclotron is of fixed energy and can accelerate protons to 62.5 MeV, which gives a beam penetration of 27 mm. This is adequate to treat 90% of ocular melanoma lesions (Goiten et al. 1983). The study is a joint project between the Douglas Cyclotron Centre, Merseyside, St. Bartholomew's Hospital and Moorfields Eye Hospital, London, and St. Paul's Eye Hospital, Liverpool. The project has the following objectives (J. Hungerford 1986, personal communication):

To evaluate the effectiveness of proton irradiation.

To establish the nature and incidence of complications from proton therapy of uveal melanoma.

To establish the long-term effect of proton therapy on visual acuity.

To compare in a randomized study proton therapy versus ^{125}I scleral applicator treatment for small melanomas adjacent to the macula or optic nerve.

To study the relationship between local control and distant metastases for proton therapy.

To compare as part of a randomized study the mortality following proton therapy with that following enucleation for medium-sized and large melanomas.

Deeper tumour sites treated with protons (160–250 MeV) continue to be studied. Randomized trials have been started at Harvard for prostatic cancers and chordomas or low-grade chondrosarcomas at the base of the skull or cervical spine. The good results from non-randomized studies, restricted machine availability, patient-specific treatment techniques and the advantages of proton treatment being judged sufficiently great to preclude a randomized study, have created problems in setting-up the randomized trials that are clearly needed (Goiten and Munzenrider 1984).

Neutron Therapy

Since the early clinical studies of Stone in the 1940s considerable energy has been expended on investigating the potential of neutron therapy both in the laboratory and in the clinic. These studies have been conducted predominantly in the United States, the United Kingdom and the rest of Europe.

Three major changes have occurred in neutron therapy during the past 10 years:

1. A move from non-hospital-based cyclotrons to hospital-based cyclotrons capable of generating neutron beams with sufficient output.
2. A change from fixed beam to isocentric machines with internal collimation.
3. A change from low-energy to high-energy neutron beams.

The Development of Neutron Therapy

Following the 1948 Janeway lecture by Stone (1948) interest in neutron therapy in the United States initially declined. When further radiobiological experiments were performed at the MRC Cyclotron Unit, Hammersmith Hospital, London, a more tolerable clinical regimen evolved using the Hammersmith 15-MeV deuteron on beryllium neutron beam with a mean energy of 7.5 MeV. Despite the inadequacies of this beam in terms of dose distribution and collimation difficulties as compared with megavoltage X-rays, very encouraging results were reported (Catterall and Bewley 1979) and many centres worldwide were stimulated to investigate neutron therapy further.

Initially attempts were made to use fractionation regimes similar to those commonly used for photons. Using such 5-day-a-week regimes, however, problems were encountered with morbidity and access to machines not based in hospitals. For this reason, as well as the desire to continue treating 5 days a week, a schedule of three fractions of photons and two fractions of neutrons per week was introduced. Most of the American experience, therefore, was with this mixed-beam therapy up until 4 or 5 years ago. Many American facilities were not hospital-based and there was also a wide range of energies available (Peters et al. 1985). In Europe all but a few studies have been uncontrolled phase I/II studies (Wambersie and Batterman 1985). In general neutrons alone rather than mixed beams have been used. In the United Kingdom mixed-beam therapy has not been used and there has been a tradition of controlled clinical studies using hospital-based facilities with gradually increasing technical flexibility.

The cell biologist, as opposed to the biophysicist, may consider that the real targets for radiation therapy are the hypoxic cells, less radioresponsive cells and clonogenic cells present in tumours. It is the density and distribution of these related to the flux of the energy deposition of the treating beam that determine the degree of cell kill with each fraction of radiotherapy and thus influence the probability of effective local control. The physical characteristics of the modern high-energy, hospital-based, isocentric and internally collimated neutron facilities allow the flux of neutron beam energy in the tumour to be deposited with the same precision as that of a modern linear accelerator. In addition, however, the biological characteristics of these beams themselves may be different because of the higher energy of the fast neutrons. In contrast to the lower-energy beams available at Hammersmith and Edinburgh, neutrons produced by protons of 40 MeV and upwards, as in the facilities at Seattle and Clatterbridge, have a lower LET. The density of deposition of ionization per neutron track length is lower than with less energetic machines and the track length of the recoil protons themselves is longer. These microdosimetric differences may prove to be important in terms of the therapeutic ratio of high-energy fast neutron beams with regard to tumour control and normal tissue late radiation damage (Wambersie et al. 1986).

Current Status of Clinical Studies with Neutrons

A wide variety of tumours have now been treated with fast neutron therapy of various energies. There are, however, relatively few controlled clinical studies and important questions regarding the role of fast neutrons in radiotherapy remain to be answered. In general most experience has been gained with the more accessible tumours such as squamous cell head and neck cancer, salivary gland tumours, sarcomas of the limbs and melanomas. Pelvic and intrathoracic tumours have been less accessible to treatment with the lower energy beams and have only become accessible using precisely deployed high-energy isocentric machines in the past few years.

Squamous Cell Head and Neck Cancer

Early results obtained at the MRC Cyclotron Unit, Hammersmith Hospital, were very encouraging, although the local control rates in the photon-treated control group were acknowledged to have been disappointingly low. Many of these tumours were very advanced however. In view of the Hammersmith results a cyclotron of similar energy beam, this time with an isocentric facility, was installed at Edinburgh, with the intention of repeating the controlled study performed at Hammersmith. A restrospective analysis (Medical Research Council Neutron Therapy Working Group 1986) of squamous cell head and neck cancer treated in the controlled studies in both centres, however, revealed many differences between the two studies. Stage of disease was more advanced in the Hammersmith series and technical differences in the application of therapy were noted in that more wedged fields and three-field plans were used in the Edinburgh series. In addition different fractionation schedules were used in the two centres. It is of interest that whilst much of the American experience was with mixed-beam therapy, in a small series of patients in Seattle who were treated with neutrons alone or photons (Griffin et al. 1984) similar results have been obtained to those in the Hammersmith study. Other studies reviewed by Peters et al. (1986) have not shown a significant benefit for neutron therapy.

Salivary Gland Tumours

The efficacy of fast neutrons against malignant salivary gland tumours was first demonstrated at Hammersmith in a non-randomized series. In the most recent review (Catterall and Errington 1987) of these results local control and 5-year survival were 72% and 50% respectively. The facial nerve was not damaged by neutron therapy and 70% of patients treated with neutrons alone regained or maintained facial nerve function. These results have been confirmed by those from other neutron treatment facilities. A recently completed RTOG controlled study has shown a 78% tumour clearance rate using neutrons as compared with 33% after photon radiotherapy, with local control values of 53% and 33% respectively at 12 months (T. G. Griffin, personal communication). This study has now been closed although the results are only just approaching statistical significance. These results are in keeping with those from the previous non-randomized studies and are sufficiently impressive to allow the decision not to conduct a randomized study of neutrons versus photons for salivary gland tumours using the Clatterbridge cyclotron, but to offer neutron therapy to all patients with locally advanced malignant salivary gland tumours. The clinical problem for the future is to establish the efficacy of neutrons versus surgery for these tumours, rather than to define the optimal radiotherapeutic modality.

Paranasal Sinus Tumours

Paranasal sinus tumours are relatively rare, providing insufficient numbers for randomized controlled studies. In general they are not considered easy to control by conventional photon radiotherapy. Table 13.2 shows a non-randomized series of patients treated at Hammersmith Hospital with encouraging local control rates (Errington 1986). The number of late radiation complications in normal tissues should, however, also be noted. From this and many other studies it appears that such a complication rate may be the unavoidable price of attempting to achieve local control in such advanced tumours using low-energy fast neutrons. Such complications may be considered as contributing to a poor quality of life, but it should be remembered that uncontrolled local disease may well produce a poorer quality of life which deteriorates at a more rapid rate than that due to the late radiation tissue morbidity of low-energy fast neutron therapy. Late effects must also be set against the morbidity of radical surgery in such patients.

Table 13.2. Results of treatment with 7.5-MeV neutrons for advanced tumours of paranasal sinuses: histological types, responses and complications

Histological type	Number	No. regressing completely	No. recurring	No. with complications
Squamous	17	14	3	3
Adenoid cystic	11	10	4	4
Adenocarcinoma	8	6	—	1
Transitional cell	5	5	1	3
Undifferentiated	1	1	—	—
Malignant melanoma	1	1	—	—
Totals	43	37 (86%)	8 (18%)	[a]10 (23%)

From Errington (1986).
[a] Two of these from eight patients who had received previous photon radiotherapy.

Adenoid Cystic Carcinoma

Adenoid cystic carcinomas generally respond well initially to conventional radiotherapy but the local recurrence rate is high (Vikram et al. 1984). In a non-randomized series of 36 patients with adenoid cystic carcinoma treated with neutrons complete regression occurred in 97% with a local recurrence rate of only 9%. The median survival of the whole group was 54 months (Catterall et al. 1987). These results support the view that neutron therapy may well be the treatment of choice for non-resectable or recurrent non-squamous cell cancer of the head and neck (Cohen et al. 1985).

Fixed Cervical Nodes

Neutrons alone and also mixed-beam therapy have been employed in the treatment of advanced head and neck cancers with fixed cervical nodes. Both the uncontrolled studies of neutrons alone and a controlled study in the United States comparing mixed-beam therapy with photons (Griffin et al. 1983) suggest that there may be a role for fast neutron therapy here.

Malignant Melanoma

Malignant melanoma is often resistant to conventional photon irradiation. This has been related to the broad shoulder of the survival curves for melanoma cells and has led to alternative strategies to photon therapy being suggested (Hornsey 1978) and subsequently shown to produce high response rates in clinical practice (Overgaard et al. 1985).

In the largest series of melanomas treated by fast neutrons (Blake et al. 1985) 87 tumour sites were treated. There was complete regression in 71% and later recurrence in only 9%, giving an overall local control rate of 62%. All the tumours were locally advanced and good palliation was achieved. Fast neutron therapy is therefore an alternative in the treatment of some melanomas but has yet to be compared by a randomized study with photon therapy using high-dose per fraction schedules.

Sarcomas

Cohen et al. (1985) reviewed the international experience of neutron therapy for locally advanced non-resectable sarcomas of bone and soft tissue, and reported 59% and 62% long-term control respectively for these radioresistant tumours. Schmitt et al. (1986) have treated 191 patients with locally advanced soft tissue sarcomas with neutrons or photons followed by a reduced-volume neutron boost. In neutron treated patients the local control rate was 76% and survival best in patients with low-grade (grade 1) histologies (75% at 6.5 years). Pickering et al. (1987) report that 68% of gross tumours regressed completely and that local control has been maintained at 52%. The main cause of death in these patients was metastatic disease, though again the most favourable outlook was for patients with low-grade tumours (median survival of 63 months for grade 1 compared with 9 months for grade 2 and 7 months for grade 3). It can be concluded that slowly growing, locally advanced and well-differentiated (grade 1) soft tissue sarcomas are an absolute indication for neutron therapy.

Pelvic Tumours

Because low-energy neutron beams have insufficient penetration for treating pelvic tumours little experience had been gained regarding neutron therapy for these tumours until the past 5 years. An RTOG study of mixed-beam treatment of carcinoma of the cervix by Maor et al. (1986) showed no particular benefit for mixed-beam therapy compared with photon treatment alone. Treatment of rectal and bladder cancers by Batterman and Mijnheer (1986) using a DT generator has been carried out on an uncontrolled series of patients; no real benefit was found. It was noted that a similar dissociation to that observed by Stone (1948) between the intensity of early acute radiation effects and late radiation damage was apparent. Similar problems with morbidity from late radiation damage were highlighted by the Edinburgh group (Duncan et al. 1985, 1986), where control rates for bladder cancer using low-energy 7.5-MeV neutrons alone were no better than those with photons, whilst the morbidity in the neutron treated patients was considerably higher.

By using six treatment fields and comparable isodoses to those for the photon treated control patients, it was possible to attribute the morbidity to the biological effect of the neutron beam rather than its potential inferiority in terms of physical dose distribution. Griffin et al. (1986) have demonstrated, however, that the incidence of major complications is much lower when high-energy (e.g. 46–66 MeVp-Be) neutron beams are used. This question is only likely to be resolved when the current controlled randomized studies of high-energy fast neutrons versus megavoltage X-rays in pelvic sites (bladder, rectum, cervix and prostate) are completed.

Carcinoma of the Prostate

More encouraging results in the pelvis have been obtained in a mixed-beam study in the United States in which patients randomized to receive neutron therapy showed not only improved local control but also better survival than patients randomized to be treated with photons alone (Laramore et al. 1985). The result was initially thought to be due to the selection process by which patients who had not completed the relevant radiation protocols appropriately were not included in the final analysis. However, even when all the patients irrespective of any protocol violation were included the mixed-beam group appeared to have a significantly higher survival than the control group. A randomized study of neutrons alone versus photons, using high-energy neutrons produced by hospital-based cyclotrons, has been started by the RTOG in the United States.

The Therapeutic Index for High- and Low-Energy Fast Neutrons

In assessing the effect of neutron therapy in controlled studies the therapeutic index for each arm should probably be expressed as:

$$\frac{\%\ \text{local tumour control}}{\%\ \text{late radiation damage}}$$

This ratio should be scored at 1–2 years when the majority of tumours which are not controlled will have recurred and late radiation damage will have become apparent in surviving patients with local tumour control.

Whilst such therapeutic indices for high-energy fast neutrons are not yet available, evidence has accumulated that late radiation damage from high-energy neutrons may not be as bad as that from low-energy neutrons at doses which produce comparable acute reactions. This information was obtained in a retrospective study of facilities in the United States where results from neutron beams of a range of energies were analysed (Griffin et al. 1986). The late complications (grade IV and V toxicities) were scored on previously treated patients who had received low-, intermediate- and high-energy neutron therapy. For both head and neck and pelvic tumours late radiation complications were not as marked for patients treated by the high-energy facilities as for those receiving treatment from low-energy facilities. Prospective dose ranging studies at three different dose levels were then undertaken. These studies closed in March 1986. Only a small number of patients have yet been studied for sufficient time to observe late radiation complications, but few problems have been noted to date in those patients treated with high-energy neutrons (T. W. Griffin, 1986, personal communication). The acute reactions observed in these patients were very similar to those in patients treated radically with photon therapy. It is thus possible that the late radiation damage observed with low-energy fast neutron beams may not be such a problem with higher-energy beams.

At the Medical Research Council Douglas Cyclotron Centre at Clatterbridge Hospital radiobiological experiments were conducted using acute skin reactions and gut clone studies to compare the biological effect of the Clatterbridge beam to that used at Hammersmith Hospital in previous studies. It was predicted that at the high energy a total effective dose of 20.4 Gy in 12 fractions over 28 days would have the same biological effect as 15.6 Gy (neutron dose) in 12 fractions over 28 days from the Hammersmith 7.5-MeV beam. This figure is extremely close to that obtained for acute reactions in the American dose searching studies. These studies were also conducted using 12 fractions in 28 days.

The neutron beam at Clatterbridge has now been operating for 1 year and 53 patients have been treated. Because of the need for caution with regard to the possibility of late bowel damage a total tumour dose of 19.2 Gy in 12 fractions over 28 days given in two phases (a large and then small volume when feasible) is being used to treat locally advanced pelvic tumours.

The Future of Neutron Therapy

Work with fast neutron therapy over the past 20 years has had to be carried out in the face of formidable difficulties because of the biological effects of the low-energy beams, poor physical dose distribution and often lack of isocentric or even hospital-based facilities. In this context the skill and attention to detail that have been required to produce some noteworthy results in the face of such difficulties should be appreciated. Encouraged by these early results hospital-based high-energy neutron facilities have been developed so that further studies can be done to help define the role of neutrons in modern radiotherapeutic practice.

Modern high-energy proton-accelerating cyclotrons coupled to isocentric internally collimated treatment heads are capable of producing physical dose distributions of the same quality as those produced for photons from a modern linear accelerator. At Clatterbridge the 62.5-MeVp-Be beam is being used to conduct randomized studies of squamous cell head and neck cancer. Locally advanced malignant salivary gland tumours, paranasal sinus tumours and some previously irradiated and advanced recurrent head and neck cancers (Errington and Catterall 1986) are initially being treated with neutrons on a non-randomized basis. Non-randomized studies of neutrons for soft tissue sarcomas and melanoma are also proposed. Randomized controlled studies are in progress too with locally advanced tumours of the prostate, bladder, rectum and cervix.

These groups of tumours account for just over 50% of the current failures to control local disease. It is by improving on this figure that it is hoped particle therapy will be of benefit to patients, particularly those presenting with locally advanced tumours.

References

Battermann JJ, Mijnheer BJ (1986) The Amsterdam fast neutron therapy project: a final report. Int J Radiat Oncol Biol Phys 12:2093–2099

Bewley DK (1987) European accelerator for radiotherapy with charged ions. Lancet I:318–319

Blake PR, Catterall M, Errington RD (1985) Treatment of malignant melanoma by fast neutrons. Br J Surg 72:517–519

Castro JR, Chen GTY, Blakeley EA (1985) Current considerations in heavy charged-particle radiotherapy: a clinical research trial of the University of California Lawrence Berkely Laboratory, Northern California Oncology Group and Radiation Therapy Oncology Group. Radiat Res 104:5263–5271

Catterall M (1982a) The assessment of the results of neutron therapy. Int J Radiat Oncol Biol Phys 8:1573–1780
Catterall M (1982b) Results of neutron therapy: differences, correlations and improvements. Int J Radiat Oncol Biol Phys 8:2141–2144
Catterall M, Bewley DK (1979) Fast neutrons in the treatment of cancer. Academic Press, London
Catterall M, Errington RD (1987) The implications of improved treatment of malignant salivary gland tumours by fast neutron radiotherapy. Int J Radiat Oncol Biol Phys 13:1313–1318
Catterall M, Errington RD, Bewley DK (1987) A comparison of clinical and laboratory data on neutron therapy for locally advanced tumours. Int J Radiat Oncol Biol Phys (in press)
Cohen L, Hendrickson FR, Kurup PD et al. (1985) Clinical evaluation of neutron beam therapy. Current results and prospects, 1983. Cancer 55:10–17
Duncan W, Arnott SJ, Jack WJL et al. (1985). A report of a randomized trial of d() + Be neutrons compared with megavoltage X-ray therapy of bladder cancer. Int J Radiat Oncol Biol Phys 11:2043–2049
Duncan W, Williams JR, Kerr GR et al. (1986) An analysis of the radiation related morbidity observed in a randomised trial of neutron therapy for bladder cancer. Int J Radiat Oncol Biol Phys 12:2085–2092
Errington RD (1986) Advanced carcinoma of the paranasal sinuses treated with 7.5 MeV fast neutrons. Bull Cancer (Paris) 73:569–576
Errington RD, Catterall M (1986) Re-irradiation of advanced tumours of the head and neck with fast neutrons. Int J Radiat Oncol Biol Phys 12:191–195
Field SB, Fowler JF (1984) The biological basis for neutron therapy. J Eur Radiother 5:170–175
Goitein M, Miller T (1983) Planning proton therapy of the eye. Med Phys 10:275–283
Goitein M, Gentry R, Koehler AM (1983) Energy proton accelerator necessary for treatment of choroidal melanomas. Int J Radiat Oncol Biol Phys 9:259–260
Goitein M, Munzenrider J (1984) Proton therapy: good news and big challenges. Int J Radiat Oncol Biol Phys 10:319–320
Greiner RH, Van Essen CF, Blattmann HJ et al. (1986) Pion therapy at Swiss Insitute for Nuclear Research (SIN). Int J Radiat Oncol Biol Phys 12 [Suppl 1. Proceedings of the American Society for Therapeutic Radiology and Oncology 28th annual meeting]: 98
Griffin TW, Davis, R, Laramore GE et al. (1983) Fast neutron irradiation of metastatic cervical adenopathy. The results of a randomized RTOG study. Int J Radiat Oncol Biol Phys 9:1267–1270
Griffin TW, Davies R, Hendrickson FR et al. (1984) Fast neutron radiation therapy for unresectable squamous cell carcinoma of the head and neck: the results of a randomised RTOG study. Int J Radiat Oncol Biol Phys 10:2217–2222
Griffin TW, Pajak T, Laramore GE, Davis L (1986) Analysis of neutron radiotherapy treatment complications. Bull Cancer (Paris) 73:582–586
Hall EJ (1981) New modalities in cancer treatment: heavy charged particles. Br J Radiol 54:773–781
Hornsey S (1978) The relationship between total dose, number of fractions and fraction size in the response of malignant melanoma in patients. Br J Radiol 51:905–909
Laramore GE, Krall JM, Thomas FT et al. (1985) Fast neutron radiotherapy for locally advanced prostate cancer: results of an RTOG randomized study. Int J Radiat Oncol Biol Phys 11:1621–1667
Maor MH, Gillespie B, Peters LJ et al. (1986) Neutron therapy in cervical cancer: results of a Phase III RTOG study. Int J Radiat Oncol Biol Phys 12 [Suppl 1. Proceedings of the American Society for Therapeutic Radiology and Oncology 28th annual meeting]: 99–100
McNally NJ, Wilson GD (1986) Cell kinetics of normal and perturbed populations measured by incorporation of bromodeoxyuridine and flow cytometry. Br J Radiol 59:1015–1022
Medical Research Council Neutron Therapy Working Group (1986) A comparative review of the Hammersmith (1971–75) and Edinburgh (1977–82) neutron therapy trials of certain cancers of the oral cavity, oropharynx, larynx and hypopharynx. Br J Radiol 59:429–440
Munzenrider JE, Austin-Seymour M, Blitzer PJ et al. (1985) Proton therapy at Harvard. Strahlentherapie 161:756–763
Overgaard M, Overgaard J, Sell A (1984) Dose response for radiotherapy of recurrent, residual and primarily inoperable colorectal cancer. Radiother Oncol 1:217–225
Overgaard J, Van der Maase J, Overgaard M (1985) A randomized study comparing two high-dose per fraction radiation schedules in recurrent or metastatic malignant melanomas. Int J Radiat Oncol Biol Phys 11:1837–1839
Peters LJ, Maor MH, Laramore GE et al. (1985) Review of clinical results of fast neutron therapy in the USA. Strahlentherapie 161:731–738
Peters LJ, Maor MH, Laramore GE, Griffin TW, Hendrickson FR (1986) Review of clinical results

of fast neutron therapy in the USA. In: Maruyama Y, Beach LJ, Feola JM (eds) Californium-252 brachytherapy and fast neutron beam therapy. Nuclear science applications. Harwood Academic Publishers, London, pp 243–260

Pickering DG, Stewart JS, Rampling R, Errington RD, Stamp G, Chia Y (1987) Fast neutron therapy for soft tissue sarcoma. Int J Radiat Oncol Biol Phys 13:1489–p1495

Schmitt G, Furst G, Bamberg M (1986) The value of neutron and neutron-boost irradiation for the local control of advanced soft tissue sarcomas. Bull Cancer (Paris) 73:577–581

Seydel Gunter H (1984) Radiation sensitizers. In: Gilbert Harvey A (ed) Modern radiation oncology. Classic literature and current management, vol 2. Harper and Row, Philadelphia, pp 664–686

Stone RS (1948) Janeway memorial lecture. Neutron therapy and specific ionization. Am J Roentgenol 59:771–785

Stone RS, Larkin JC (1942) The treatment of cancer with fast neutrons. Radiobiology 39:608–620

Suit HD, Goiten M (1980) Rationale for use of charged-particle and fast-neutron beams in radiation therapy. In: Meyn RE, Withers HR (eds) Radiation biology in cancer research. Raven Press, New York, pp 547–565

Thames HD, Peters LJ, Withers HR, Fletcher GH (1983) Accelerated fractionation vs hyperfractionation: rationales for several treatments per day. Int J Radiat Oncol Biol Phys 9:127–138

Vikram B, Strong EW, Shah JP, Spiro RH (1984) Radiation therapy in adenoid cystic carcinoma. Int J Radiat Oncol Biol Phys 10:221–223

Von Essen CF, Bagshaw MM, Bush SE, Smith AR, Kligerman MM (1986) Long-term results of pion therapy at Los Alamos. Int J Radiat Oncol Biol Phys 12 [Suppl 1. Proceedings of the American Society for Therapeutic Radiology and Oncology 28th annual meeting]: 97

Wambersie A, Batterman JJ (1985) Review and evaluation of clinical results in the EORIC Heavy-Particle Therapy Group. Strahlentherapie 161:746–755

Wambersie A, Bewley DK, Lalanne CM (1986) Prospects for the application of fast neutrons in cancer therapy. Radiobiological bases and survey of the clinical data. Bull Cancer (Paris) 73:546–561

14 Radiation Sensitizers and Related Compounds: New Approaches to Modification of Radiation Response

G. Adams and I. J. Stratford

Introduction

The hypothesis that the relative radiation resistance of hypoxic tumour cells present in some human tumours can adversely affect the probability of local control in some clinical situations, remains a major research topic. There is some clinical evidence, mainly from studies using hyperbaric oxygen, that hypoxic cell resistance is a factor in the local control of some tumours. However, it is widely believed that tumour hypoxia is probably only important in some tumours; in others, reoxygenation occurring between fractions many effectively remove hypoxic cells during treatment. It is against the background of this heterogeneity with respect to the importance of hypoxia that attempts to demonstrate clinical radiation sensitization have been carried out. In retrospect, therefore, it is perhaps not surprising that the results of clinical trials with misonidazole have been generally disappointing. An urgent need is to develop methods of identifying those human tumours where hypoxic cell radioresistance is a problem. When such techniques become available the clinical efficacy of any new and improved sensitizer should be much more easily demonstrable.

This chapter describes some of the current laboratory approaches now being undertaken to develop more effective methods of hypoxic cell radiation sensitization.

Current Hypoxic Cell Sensitizers

Experimental studies with misonidazole showed that, provided a sufficiently high dose of the drug could be administered, substantial sensitization occurs in almost all solid experimental rodent and human tumours irradiated with single doses of radiation. It is reasonable, therefore, to attribute, at least in part, the failure to observe significant sensitization in most clinical trials of misonidazole to the dosage limitations imposed by the neurotoxic properties of this drug (Dische et al. 1977).

Two compounds potentially superior to misonidazole are undergoing clinical investigation. These are etanidazole and pimonidazole.

Etanidazole (SR 2508)

Laboratory studies with misonidazole and various structurally related sensitizers suggested that the neurotoxicity of compounds of the nitroimidazole type was partly related to their lipophilic properties (Brown and Workman 1980; Clarke et al. 1982). These findings led directly to the development of the misonidazole analogue, etanidazole, as a clinical hypoxic cell sensitizer (Brown et al. 1981). This drug is much less lipophilic than misonidazole and has proved to be less neurotoxic in current clinical trials (Coleman et al. 1984). While some peripheral neuropathies have been observed, they have been invariably mild, much less common, and occur at higher doses compared with the neurotoxic effects seen with misonidazole.

The sensitizing efficiency of etanidazole is similar to that of misonidazole both in vitro and in vivo. Tumour penetration appears to be generally good and this, together with the finding that the *total* tolerated dose is substantially greater than that of misonidazole, has led to optimism with regard to clinical efficacy. It is unlikely, however, that the maximum degree of sensitization theoretically achievable (i.e. an enhancement ratio similar to that of oxygen itself) will be attainable with the maximum tolerated doses of this drug.

Pimonidazole (Ro 03–8799)

The Roche drug pimonidazole represents a logical development from misonidazole (Smithen et al. 1980). It is a 2-nitroimidazole with a substituted propanolamine side chain. Sensitizers of this type containing prototropic side chains, that is possessing acid–base characteristics, are often more efficient than would be expected on purely electron-affinic grounds. Pimonidazole is slightly more efficient than misonidazole in experimental animal tumour systems (Williams et al. 1982) and is undoubtedly less toxic in patients (Saunders et al. 1984; Roberts et al. 1984). However, an unexpected and promising finding is that the drug shows a significant degree of selectivity in uptake in human tumour tissues (Roberts et al. 1984; Dische et al. 1986).

Fig. 14.1a reproduces the data of Dennis et al. (1985) showing that the sensitizing efficiency of pimonidazole in V79 cells in vitro is highly dependent on the pH of the culture medium. This effect, which is not seen with misonidazole, has been attributed to the influence of the acid–base characteristics of pimonidazole on its intracellular uptake. It is argued that entry into the cell is effected through the unprotonated form of the drug. However in the *extra*cellular fluid in hypoxic regions of tumours, the relatively lower pH allows the drug to be concentrated in its water-soluble *protonated* form. According to

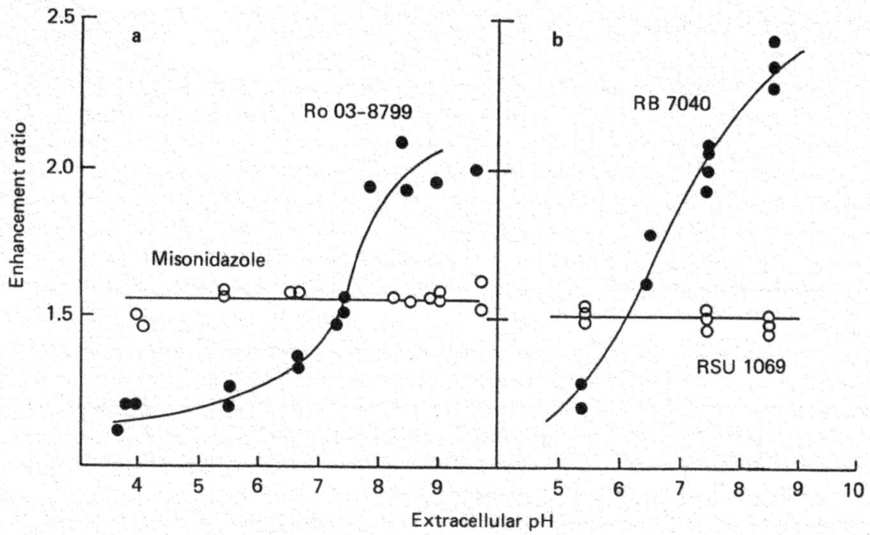

Fig. 14.1. Influence of extracellular pH on the sensitizing efficiencies of some electron affinic nitroimidazole radiation sensitizers in Chinese hamster V79 cells irradiated in vitro. **a** Data for misonidazole and Ro 03–8799 taken from Dennis et al. (1985) and Watts and Jones (1985). **b** Data for RSU 1069 and RB 7040 taken from Walling et al. (1987).

the hypothesis, intracellular levels build up as the drug diffuses into the cell in the non-protonated form (Wardman 1982). The possibility of improving tumour uptake of sensitizers by manipulating their acid–base properties is a promising avenue for the future development of new drugs.

Multiple Mechanism Approach to Hypoxic Cell Sensitization

Oxygen Mimetic Sensitization

The original impetus in the development of sensitizers for hypoxic cells lay in the search for chemical agents that could mimic oxygen in its general sensitizing action. There is now clear evidence that the oxygen effect involves more than one component, is a free radical process and occurs in all types of cells. Many chemical compounds mimic oxygen in their ability to increase the radiation sensitivity of hypoxic cells but not aerobic cells. It is well established that the link between oxygen and these chemical sensitizers is their common ability to act as electronic acceptors (Adams and Dewey 1963; Raleigh et al. 1973; Adams et al. 1976). Measurements of electron affinity have provided a basis therefore for the design and development of new sensitizers (Adams et al. 1979). It is appropriate to regard these sensitizers as oxygen mimetic since, like oxygen, they act by more than one free radical component.

Bioreductive Activation

Many electron-affinic compounds containing nitro groups are much more toxic to hypoxic cells even in the absence of radiation than they are under oxic conditions. For some compounds this differential effect can be substantial. There is now little doubt that the enhanced hypoxic cytotoxic response is caused by highly toxic substances formed from the parent compound via metabolic bioreduction occurring in the hypoxic cells (see Alexander 1986). The mechanism is probably similar in some respects to that whereby some sensitizers such as metronidazole can act as powerful antibiotics for various pathogenic anaerobes. Fig. 14.2 shows some data for misonidazole and the powerful bioreductive drug RSU 1069. For both drugs there is a marked increase in the toxic response under hypoxic compared with oxic conditions.

Neither the biochemical nature of the bioreductive process nor the identity of the reduced products is known with precision, although the evidence is substantial that the process is initiated through chemical reduction of the nitro group on the imidazole ring (see Brown 1986).

Bioreductive activation appears to be a general property of many radiation sensitizers, particularly nitroimidazoles. It should be noted, however, that while bioreductive activation can contribute under some conditions to overall radiation response, such effects are separate from, and additional to, the main mechanism of oxygen mimetic sensitization involving free radical processes.

Bioreductive activation may be a factor in the ability of some radiation sensitizers apparently to act as inhibitors of repair processes. In some tumour

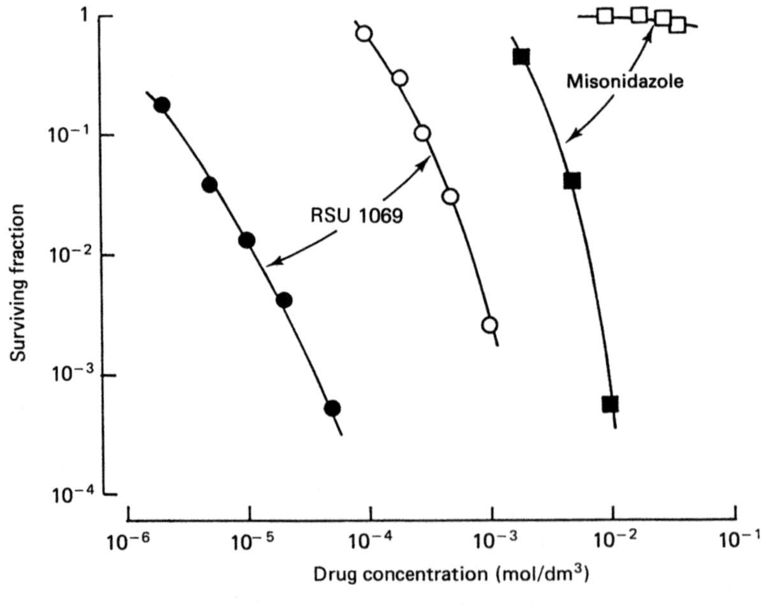

Fig. 14.2. Toxicity of misonidazole and RSU 1069 to Chinese hamster V79 cells: *open symbols*, in air; *filled symbols*, in nitrogen. Cells in confluent cultures were exposed to drugs for 3 hours at 37 °C. (After Stratford et al. 1986.)

systems, where tumour response in vivo can be assayed in vitro by clonogenic assay, the degree of sensitization by misonidazole can be enhanced if excision of the tumour post-irradiation is delayed for 24 hours. This effect has been interpreted as an inhibitory effect of misonidazole on repair of potentially lethal damage (Guichard et al. 1979; Nakagawa et al. 1983).

Effects noted by Hall and Biaglow (1977). Wong et al. (1978), Stratford et al. (1980) and Chapman et al. (1982) may also relate to bioreductive activation. In particular, Wong et al. (1978) observed that cells exposed in hypoxia to sublethal amounts of misonidazole, then washed free of the drug and irradiated in either oxygen or nitrogen, exhibited complete suppression of the shoulder region of the radiation survival curve. Since the effect is only observed when the pre-irradiation incubation is carried out in hypoxia, bioreductive activation of the drug is implicated.

Thiol Suppression

There is unequivocal evidence that alterations in the intracellular levels of glutathione (GSH) influence cellular radiosensitivity both in vitro and in vivo. Reduction in GSH levels can be brought about by treatment with thiol binding agents such as diamide (Harris 1979) and diethylmaleate (Bump et al. 1982). Alternatively, inhibition of GSH biosynthesis by the inhibitor of γ-glutamyl cysteine synthetase, buthionine sulphoximine (BSO), can also lead to prolonged reduction of intracellular GSH levels (Griffith and Meister 1979). An increase in radiosensitivity occurs under both hypoxic and aerobic conditions but not necessarily to the same extent. When hypoxic cell sensitizers are present in cells with reduced GSH levels, the efficiency of hypoxic cell sensitization is considerably increased. For a detailed discussion of the mechanisms involved see Biaglow et al. (1983).

Dual Function Radiosensitizers

There is current interest in the "dual-function" class of nitroheterocyclic sensitizers, so called because of the incorporation into the chemical structure of both radiation sensitizing and alkylating properties. Some of these agents are particularly potent radiation sensitizers and bioreductive drugs. The lead compound in this series is the misonidazole analogue RSU 1069, which combines a 2-nitroimidazole structure with a monofunctional aziridine group in the side chain (Adams et al. 1984a,b).

Fig. 14.2 shows the differential cytotoxic properties of RSU 1069 in oxic and hypoxic cultures of Chinese hamster V79 cells (Stratford et al. 1986). The cells were incubated for 3 hours under either oxic or hypoxic conditions in the presence of various concentrations of the drug, resuspended, washed free of drug and aliquots plated for colony formation. The drug is considerably more cytotoxic than misonidazole under both oxic and hypoxic conditions, but the hypoxic differential is particularly marked for RSU 1069. Under the conditions of the experiment the compound is more cytotoxic than misonidazole by at least two orders of magnitude.

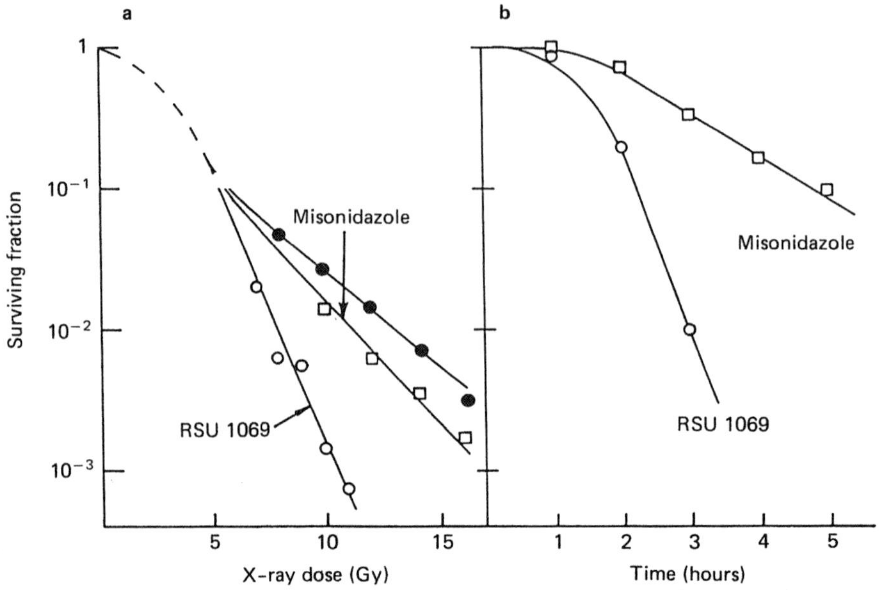

Fig. 14.3. Comparison of the radiation sensitizing efficiencies and hypoxic cytotoxic efficiencies of misonidazole and RSU 1069. **a** Radiation sensitization of the KHT tumour by 100 mg/kg misonidazole and 80 mg/kg RSU 1069. *Filled symbols*, radiation alone. **b** Cytotoxic effect of 5mM misonidazole and 0.05 mM RSU 1069 in hypoxic cultures of V79 cells exposed to each drug for up to 5 hours.

Fig. 14.3 compares both the radiation sensitizing and cytotoxic properties of misonidazole and RSU 1069. Fig. 14.3a refers to sensitization of the KHT tumour impanted subcutaneously on the backs of C3H mice. The tumours were irradiated 1 hour after intraperitoneal administration of the drug dissolved in saline, excised 24 hours later and disaggregated for clonogenic assay of surviving cells. The KHT tumour has about 10% hypoxic fraction. It would be expected, therefore, that the control survival curve would contain components representing the oxic and hypoxic subpopulations. The data points correspond to changes in the survival of the fraction of hypoxic cells in the tumour. Clearly, RSU 1069 is considerably more efficient than misonidazole in sensitizing this tumour to radiation, a result in line with the greater sensitizing activity of the drug found previously in vitro (Adams et al. 1984a). Fig. 14.3b compares directly the hypoxic cytotoxic activity of the two drugs in hypoxic cultures of V79 cells exposed to each drug for up to 5 hours. Despite the 100-fold lower molar concentration of RSU 1069, the drug is much more cytotoxic than misonidazole.

Differential Uptake

The compound pimonidazole (Ro 03–8799) is an example of a sensitizer where increased uptake by tumour relative to blood results in an increase in therapeutic efficiency. As discussed above this effect has been attributed to the acid–base properties of the side chain in the chemical structure of the drug. It is important to determine, therefore, whether this behaviour can be exploited in the design of improved sensitizers. Promising results have been obtained with analogues of RSU 1069 in support of this.

The compound RB 7040 is an analogue of RSU 1069 in which the aziridine side chain is tetra-methylated. This substitution increases the pK_a value from 6.0 (for RSU 1069) to 8.45 – a value fairly close to that of 8.9 for pimonidazole (Smithen et al. 1980). Fig. 14.1b shows that the sensitizing efficiency of RB 7040, like that of pimonidazole, is highly dependent on the pH of the external culture medium. However measurements of its intracellular uptake showed that the apparent pH dependence of sensitizing efficiency is due entirely to the influence of the pH of the medium on the intracellular/extracellular distribution of the sensitizer (Walling et al. 1987). The property of differential uptake may be a general characteristic of weakly basic nitroimidazole sensitizers ($pK_a > 7$). If so, the pK_a of the compound may be a generally useful property to manipulate in the design of improved sensitizers and bioreductive drugs.

Manipulation of Tumour Hypoxia for Therapeutic Gain

Haemoglobin Status

It is well known that anaemia is a poor prognostic indicator for local control in the radiotherapy of some tumours (Bush et al. 1978). The possible criticality of haemoglobin status within the normal range has been indicated by the data of Overgaard et al. (1986, 1987) obtained in a large prospective randomized trial of radiotherapy in the treatment of some pharyngeal tumours. An improved response rate was reported for patients with haemoglobin values towards the upper end of the normal range compared with that for patients with lower values. No patient in either group was *clinically* anaemic requiring transfusion before treatment.

A surprising finding with regard to haemoglobin status emerged from a re-examination (Dische et al. 1983) of the results of treatment with hyperbaric oxygen (HBO) in the radiotherapy of advanced cervical cancer, including data from the Medical Research Council's multicentre trial (Watson et al. 1978). Thirty-nine patients required transfusion before treatment, of whom 16 were treated in HBO and 23 in air. Despite transfusion the air group showed a rather poor response whereas the transfused patients treated in oxygen showed a response that appeared to be even better than that for the total group of patients treated in oxygen.

This suggestion that prior anaemia may have a *beneficial* effect on treatment in oxygen or with other sensitizers has prompted much research on anaemia and its effect on the radiation response of experimental tumours (Hill et al. 1971; Hewitt and Blake 1971; Hirst 1986; Hirst et al. 1984; Adams et al. 1986).

Modification of Haemoglobin–Oxygen Affinity

Temporary reduction in the levels of tumour oxygenation can be achieved by deliberate displacement of the oxygen–haemoglobin association equilibrium. An increase in the affinity of haemoglobin for oxygen (i.e. a left shift in the curve) can be brought about by small changes in temperature or pH or by decreasing erythrocyte concentrations of 2,3-diphosphoglycerate, an allosteric effector for the binding of oxygen to haemoglobin. The Burroughs Wellcome compound BW12C under development as an anti-sickling agent is a potent left-shifter (Beddel et al. 1984). Reduction in levels of free oxygen in blood can induce radioprotection in some normal tissues, apparently induces close to 100% hypoxia in the Lewis Lung tumour and results in necrosis in murine lymph nodes infiltrated with a rapidly growing experimental T cell lymphoma (Adams et al. 1986).

Induction of Tumour Hypoxia by Vasoactive Drugs

There is much experimental evidence that even small tumours are considerably restricted in their oxygen supply compared with most normal tissues. Clinical evidence is accumulating that oxygen deficiency may be a feature of many human solid tumours also. The possibility that further deprivation of oxygen could induce tumour necrosis and render such tumours more susceptible to

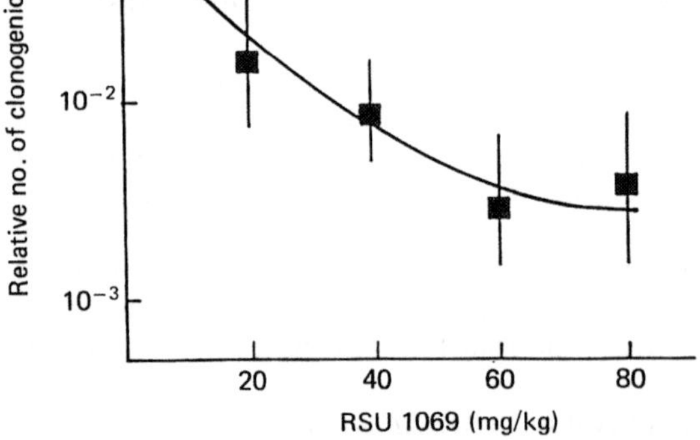

Fig. 14.4. Effect of a single dose of hydralazine on the cytotoxicity of RSU 1069 in the Lewis Lung tumour. The drug RSU 1069 was administered 15 minutes before a single intravenous dose of 5 mg/kg hydralazine. *Dashed line*, RSU 1069 alone.

subsequent therapy has led to much investigation into methods of artificially inducing tumour hypoxia.

A recent development has involved the exploitation of the "steal phenomenon" using vasoactive drugs such as the anti-hypertensive agent hydralazine (Chaplin and Acker 1987). This drug, which acts at the level of the vascular smooth muscle, induces vasodilation in normal tissues but would be expected to have less effect on the primitive vaculature in experimental tumours. Overall, blood is diverted from the tumour and thus increases the level of tumour hypoxia (Chaplin and Acker 1987; Stratford et al. 1987a). This effect can increase the efficiency of bioreductive agents, enhance the effectiveness of some anti-cancer agents such as melphalan (Chaplin and Acker 1987; Adams and Stratford 1987; Stratford et al. 1987b) and increase the efficiency of some electron-affinic sensitizers (Stratford et al. 1987a).

Fig. 14.4 illustrates the effect of hydralazine in increasing the hypoxic cytotoxic effect of RSU 1069 in the experimental Lewis Lung tumour. Normally, about 10% of the clonogenic cells of this tumour are radiobiologically hypoxic when the tumour is grown subcutaneously on the backs of C57 mice. Animals were treated either with single doses of RSU 1069 alone or with single doses of RSU 1069 followed 15 minutes later by a dose of 5 mg/kg hydralazine. The dashed line in Fig. 14.4 shows only a slight effect of RSU 1069 alone, with significant tumour cell kill only for drug doses in excess of about 40 mg/kg. In contrast, the cell kill in the hydralazine-treated mice is greatly increased due to the influence of the induced tumour hypoxia on the cytotoxic action of the bioreductive drug.

Exploitation of Induced Tumour Hypoxia for Increasing Efficiency of Radiosensitization

The data in Fig. 14.5 indicate that induction of hypoxia post-irradiation enhances the efficiency of the electron-affinic sensitizer misonidazole in the KHT tumour. Fig. 14.5a shows that in this tumour a single dose of 100 mg/kg misonidazole administered 60 minutes before a single dose of X-radiation causes only a small increase in tumour sensitivity corresponding to an enhancement ratio of about 1.3 calculated from D_0 ratios. However, when hydralazine is given post-irradiation there is a considerable increase in sensitization corresponding to an enhancement ratio of about 2.4. The data in Fig. 14.5b show that a single X-ray dose of 10 Gy reduces the surviving fraction of clonogenic cells in the KHT tumour to about 3×10^{-2}. This is decreased to 5×10^{-3} as the administered dose of misonidazole given before irradiation is increased to 500 mg/kg. In contrast, when hydralazine is administered post-irradiation the sensitization efficiency of misonidazole is enhanced by about one order of magnitude. Doses of misonidazole of 100 mg/kg or more when combined with hydralazine give the maximum enhancement ratio achievable in this tumour system with misonidazole alone (Stratford et al. 1987a).

Similar effects of hydralazine treatment have been observed in the radiation sensitization of the KHT tumour by the drug RSU 1069. The mechanism of this increase in sensitization efficiency has not yet been fully explained, although it is likely to be due, in part, to the cytotoxic effect of the bioreductive agent in a

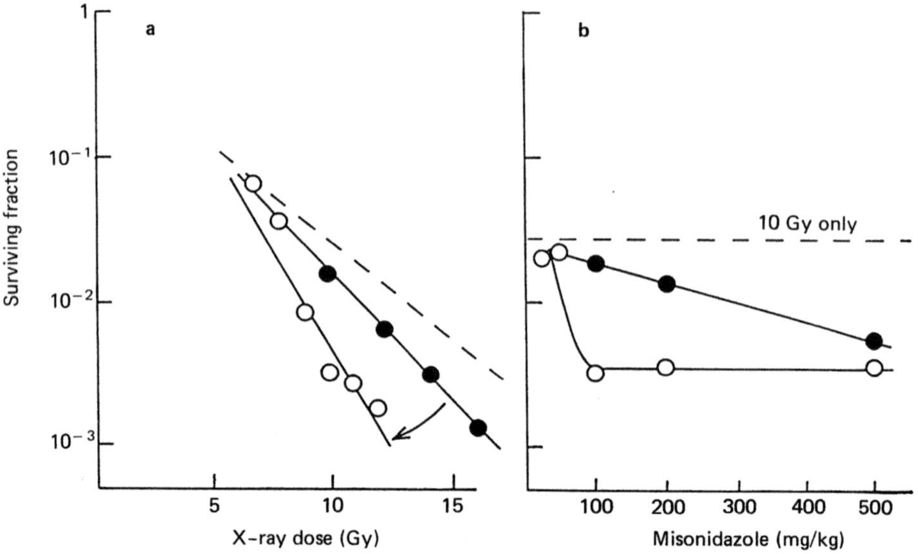

Fig. 14.5. Response of the KHT sarcoma to single doses of X-rays. **a** Dependence on radiation dose (100 mg/kg misonidazole ± hydralazine). **b** Dependence on misonidazole concentration. *Broken line*, control tumours; *filled circles*, tumours from mice treated only with misonidazole 60 minutes prior to irradiation; *open circles*, as above but with 5 mg/kg hydralazine administered intravenously immediately after irradiation.

tumour cell population rendered close to 100% hypoxic by the hydralazine treatment.

References

Adams GE, Dewey DL (1963) Hydrated electrons and radiobiological sensitization. Biochem Biophys Res Comm 12:473–477

Adams GE, Stratford IJ (1987) Sensitization of anti-cancer agents. In: Dubois JB, Joyeux H, Serrou B (eds) Colorectal, ovarian, liver cancer. Colloq INSERM 150:151–162

Adams GE, Flockhart IR, Smithen CE, Stratford IJ, Wardman P, Watts ME (1976) Electron-affinic sensitization. 7. A correlation between structure, one-electron reduction potentials and efficiencies of nitroimidazoles as hypoxic cell radiosensitizers. Radiat Res 67:9–20

Adams GE, Clarke ED, Flockhart IR et al. (1979) Structure–activity relationships in the development of hypoxic cell radiosensitizers. 1. Sensitizing efficiency. Int J Radiat Biol 35:133–150

Adams GE, Ahmed I, Sheldon PW, Stratford IJ (1984a) Radiation sensitization and chemopotentiation: RSU 1069, a compound more efficient than misonidazole in vitro and in vivo. Br J Cancer 49:571–578

Adams GE, Ahmed I, Sheldon PW, Stratford IJ (1984b) RSU 1069, a 2-nitroimidazole containing an alkylating group: high efficiency as a radio- and chemosensitizer in vitro and in vivo. Int J Radiat Oncol Biol Phys 10:1653–1656

Adams GE, Barnes D, Loutit J et al. (1986) Induction of hypoxia in normal and malignant tissues by changing the oxygen affinity of haemoglobin: implications for therapy. Int J Radiat Oncol Biol Phys 12:1299–1302

Alexander P (1986) Bioreduction in the activation of drugs. Biochem Pharmacol 35:1–122
Beddell CR, Goodford PJ, Kneen G, White RD, Wilkinson S, Wootton R (1984) Substituted benzaldehydes designed to increase the oxygen affinity of human haemoglobin and inhibit the sickling of sickle erythrocytes. Br J Pharmacol 82:397–407
Biaglow JE, Varnes ME, Clark EP, Epp ER (1983) The role of thiols in cellular response to radiation and drugs. Radiat Res 95:437–455
Brown JM (ed) (1986) Chemical modifiers of cancer treatment. Int J Radiat Oncol Biol Phys 12:1019–1545
Brown JM, Workman P (1980) Partition coefficient as a guide to the development of radiosensitizers which are less toxic than misonidazole. Radiat Res 82:171–190
Brown JM, Yu NY, Brown DM, Lee W (1981) SR 2508: a 2-nitroimidazole amide which should be superior to misonidazole as a radiosensitizer for clinical use. Int J Radiat Oncol Biol Phys 7:695–703
Bump EA, Yu NN, Brown JM (1982) The use of drugs which deplete intracellular glutathione as radiosensitizers of hypoxic tumour cells *in vivo*. Int J Radiat Oncol Biol Phys 8:439–442
Bush RS, Jenkin RDT, Allt WEC et al. (1978) Definitive evidence for hypoxic cells influencing curve in cancer therapy. Br J Cancer 37 [III]: 255–258
Chaplin DJ, Acker B (1987) Potentiation of RSU 1069 tumour cytotoxicity by hydralazine: a new approach to selective therapy. Int J Radiat Oncol Biol Phys 13 (in press)
Chapman JD, Ngan-Lee J, Stobbe CC, Meeker B (1982) Radiation-induced metabolism induced reactions of hypoxic sensitizers with cellular macromolecules. In Breccia A, Rimondi C, Adams GE (eds) Advanced topics on radiation sensitizers of hypoxic cells. Plenum, New York, pp 91–100 (Nato advanced study series, vol A43)
Clarke C, Dawson KB, Sheldon PW, Ahmed I (1982) Neurotoxicity of radiation sensitizers in the mouse. Int J Radiat Oncol Biol Phys 8:787–790
Coleman CN, Urtason RC, Wasserman TH et al. (1984) Initial report of the phase I trial of the hypoxic cell radiosensitizer SR 2508. Int J Radiat Oncol Biol Phys 10:1749–1753
Dennis MF, Stratford MRL, Wardman P, Watts ME (1985) Cellular uptake of misonidazole and analogues with acidic or basic functions. Int J Radiat Biol 47:629–643
Dische S, Saunders MI, Lee ME, Adams GE, Flockhart IR (1977) Clinical testing of the radiosensitizer Ro 07–0582: experience with multiple doses. Br J Cancer 35:567–579
Dische S, Anderson PJ, Sealy R, Watson ER (1983) Carcinoma of the cervix: anaemia, radiotherapy and hyperbaric oxygen. Br J Radiol 56:251–255
Dische S, Saunders MI, Bennett MH et al. (1986) A comparison of the tumour concentrations obtainable with misonidazole and Ro 03–8799. Br J Radiol 59:911–917
Griffith OW, Meister A (1979) Potent and specific inhibition of gluthathione synthesis by buthionine sulphoximine (*S-n*-butyl homocysteine sulphoximine). J Biol Chem 254:7558–7560
Guichard M, de Langen-Omri F, Malaise EP (1979) Influence of misonidazole on the radiosensitivity of a human melanoma in nude mice: time dependent increase in surviving fraction. Int J Radiat Oncol Biol Phys 5:487–489
Hall EJ, Biaglow JE (1977) Ro–07–0582 as a radiosensitizer and cytotoxic agent. Int J Radiat Oncol Biol Phys 2:521–530
Harris JW (1979) Mammalian cell studies with Diamide. Pharmacol Ther 7:375–384
Hewitt HB, Blake E (1971) Effect of induced host anaemia on the viability and radiosensitivity of murine malignant cells *in vivo*. Br J Cancer 25:323–336
Hill RP, Bush RS, Yeung P (1971) The effect of anaemia on the fraction of hypoxic cells in an experimental tumour. Br J Radiol 44:299–304
Hirst DG (1986) Anaemia: a problem or opportunity in radiotherapy? Int J Radiat Oncol Biol Phys 12:2009–2017
Hirst DG, Hazelhurst JL, Brown JM (1984) The effect of alterations in haematocrit on tumour sensitivity to X-rays. Int J Radiat Biol 46:345–354
Nakagawa K, Tsunemoto H, Watanabe I (1983) Effect of misonidazole and the radiosensitivity and repair of potentially lethal damage of L5178Y Ascites Tumour Cells. Curr J Cancer Clin Oncol 19:527–532
Overgaard J, Hansen SH, Jorgensen K, Hjelm M (1986) Primary radiotherapy of larynx and pharynx carcinoma – an analysis of some factors affecting local control and survival. Int J Radiat Oncol Biol Phys 12:515–521
Overgaard J, Hansen SH, Anderson AP et al. (1987) Misonidazole as an adjunct to radiotherapy in the treatment of invasive carcinoma of the larynx and pharynx. Proceedings of the 3rd international meeting on progress in radio-oncology, Vienna, Austria, March 1985. Raven Press, New York pp 137–147

Raleigh JA, Chapman JD, Borsa J, Kremers W, Reuvers AP (1973) Radiosensitization of mammalian cells by *p*-nitroacetophenone. 3. Effectiveness of nitrobenzene analogues. Int J Radiat Biol 23:377–387

Roberts JT, Bleehen NM, Workman P, Walton MI (1984) A phase I study of the hypoxic cell radiosensitizer Ro–03–8799. Int J Radiat Oncol Biol Phys 10:1755–1758

Saunders MI, Anderson PJ, Bennett MH et al. (1984) The clinical testing of Ro 03–8799: pharmacokinetics, toxicology, tissue and tumour concentrations. Int J Radiat Oncol Biol Phys 10:1759–1763

Smithen CE, Clarke ED, Dale JA et al. (1980) Novel (nitro-l-imidazolyl)-alkanolamines as potential radiosensitizers with improved therapeutic properties. In: Brady LW (ed) Radiation sensitizers: their use in the clinical management of cancer. Masson, New York, pp 22–32

Stratford IJ, Adams GE, Horsman MR et al. (1980) The interaction of misonidazole with radiation, chemotherapy agents, or heat: a preliminary report. Cancer Clin Trials 3:231–236

Stratford IJ, O'Neill P, Sheldon PW, Silver ARJ, Walling JM, Adams GE (1986) RSU 1069, a nitroimidazole containing an aziridine group: bioreduction greatly increases cytotoxicity under hypoxic conditions. Biochem Pharmacol 36:105–109

Stratford IJ, Adams GE, Bowler J, Embling PW, Godden J, Howells N (1987a) Induction of tumour hypoxia post-irradiation: a method for increasing the sensitizing efficiency of misonidazole and RSU 1069 in vivo. Int J Radiat Biol (submitted)

Stratford IJ, Adams GE, Embling P, Godden J, Nolan J, Timpson N (1987b) Potentiation of the anti-tumour effect of melphalan by the vaso-active drug, hydralazine. Br J Cancer (submitted)

Walling JM, Stratford IJ, Stephens M, Adams GE (1987) Dual function radiation sensitizers and bioreductive drugs: factors affecting cellular uptake and sensitizing efficiency in analogues of RSU 1069. Int J Radiat Biol (submitted)

Wardman P (1982) Molecular structure and biological activity of hypoxic cell radiosensitizers and hypoxia specific cytotoxins. In: Breccia A, Rimondi C, Adams GE (eds) Advanced topics on radiosensitizers of hypoxic cells. Plenum, New York, pp 49–76 (NATO advanced study institutes series A43)

Watson ER, Halnan KE, Dische S et al. (1978) Hyperbaric oxygen and radiotherapy: a Medical Research Council trial in carcinoma of the cervix. Br J Radiol 51:879–887

Watts ME, Jones NR (1985) The effect of extracellular pH on radiosensitization by misonidazole and acidic or basic analogues. Int J Radiat Biol 47:645–653

Williams MV, Denekamp J, Minchinton AI, Stratford MRL (1982) In vivo assessment of basic 2-nitroimidazole radiosensitizers. Br J Cancer 46:127–137

Wong TW, Whitmore GF, Gulyas S (1978) Studies on the toxicity and radiosensitizing ability of misonidazole under conditions of prolonged incubation. Radiat Res 75:541–555

15 Radiation Sensitizers in Clinical Radiotherapy

S. Dische

Introduction

A radiosensitizer should enhance radiation effects in tumour but not those in normal tissue. Many different approaches have been made towards such sensitization which include the use of hyperthermia, cytotoxic drugs and hypoxic cell radiosensitizers.

So far there is no evidence that hyperthermia and cytotoxic drugs give, with radiotherapy, anything other than an additive effect (Bleehen 1983; Steel 1983). Both hyperthermia (Chaps. 17 and 18) and a combination of cytotoxic chemotherapy with radiotherapy (Chap. 16) are discussed elsewhere in this volume.

Radiosensitization of the Hypoxic Tumour Cell

The concept of the radioresistant hypoxic tumour cell has attracted radiobiologists and radiotherapists for over 50 years, for it appears to be a means of achieving sensitization of tumour without enhancement of normal tissue effect. It was in the 1930s that Mottram provided evidence that oxygen was important as a radiosensitizer and it was he who first suggested that deprivation of oxygen might be an important cause of radiation resistance clinically (Mottram 1935). It was his colleague, Gray, who later continued the work and brought it to wide attention. In a paper published by Gray and his group in 1953 Scott showed how, in an animal tumour system, hyperbaric oxygen appeared to overcome the radioresistance of hypoxic cells and give improved tumour control (Gray et al. 1953).

Churchill-Davidson pioneered the use of hyperbaric oxygen in clinical radiotherapy (Churchill-Davidson et al. 1955). The enthusiasm of the initial case reports gave way to some scepticism and then to randomized controlled clinical trials (Table 15.1).

Table 15.1. Hyperbaric oxygen: randomized controlled trials

Therapeutic benefit	3
Margin in favour	6
No difference	6
Total	15

In the context of other cancer trials the results are impressive. However, there was no doubt that the benefits achieved were at the expense of an increased effect in normal tissues. This was due to the very high elevation of oxygen concentrations above those normally present and the small but definite associated increase in radiation effect. This, combined with the failure to improve results in some large studies and the complexity of the technique, led to its abandonment in favour of other methods of achieving hypoxic cell sensitization.

Chemical Hypoxic Cell Radiosensitizers

Adams and his colleagues working in the Gray Laboratory at Mount Vernon Hospital, Northwood, in the 1960s explored the possibility that chemical agents might restore the radiosensitivity of an oxygen-deprived cell. Although such agents might not prove such potent sensitizers as oxygen they might freely diffuse through the tissues and unlike oxygen not be metabolized; in this way they would reach the hypoxic cells in high concentration. Soon compounds were shown to be effective under laboratory conditions, but they were too toxic or too insoluble for clinical use (Adams and Dewey 1963).

In 1973 metronidazole, a drug already familiar in medicine, was shown to be an effective hypoxic cell sensitizer (Foster and Willson 1973). A year later, with the demonstration of a more potent nitroimidazole, Ro 07–0582, later to be called misonidazole (Asquith et al. 1974), the way was open for the clinical testing of chemical hypoxic cell radiosensitizers.

The early work with misonidazole in man gave great encouragement. Radiosensitization of skin made artificially hypoxic was clearly demonstrated (Dische et al. 1976; Gray et al. 1976). Also, in a number of patients with multiple subcutaneous nodules of tumour an increased radiation response was seen when treatment was given in a single dose with sensitizer and the result compared with the use of radiation alone (Thomlinson et al. 1976).

The neurotoxicity which the drug caused when multiple doses were administered led to a dose limitation in clinical practice (Dische et al. 1977). Despite this the simple oral administration required and the great interest in hypoxic cell radiosensitization led to widescale clinical trials in the years 1978 to

1984. The results of 45 studies using both metronidazole and misonidazole are shown in Table 15.2. Only in eight was there any indication of improved response and only in two, both studying head and neck tumours, was there evidence for a worthwhile long-term benefit. It was concluded that both metronidazole and misonidazole, in the doses which could be given to patients, were weak sensitizers.

Table 15.2. Chemical sensitizing agents: randomized controlled trials

	Metronidazole	Misonidazole
Therapeutic benefit	0	2
Significantly improved results	2	4
Margin in favour (not significant)	0	2
No difference	4	30
Margin against (not significant)	0	1
Adverse response	0	0
Total	6	39

Clinical Evidence for Radiobiological Hypoxia

If radiobiological hypoxia is a problem in the clinic then the obvious step forward is to produce a hypoxic cell sensitizer more potent than misonidazole. However, the general failure to benefit from the large clinical effort of testing misonidazole has raised considerable doubts as to whether clinical benefit could ever be gained by the radiosensitization of resistant hypoxic tumour cells. There are several pieces of evidence suggesting that hypoxia is indeed a problem.

The radioprotective effect of hypoxia seen in the laboratory, in vitro and in vivo can certainly be demonstrated in man and most clearly in limbs made hypoxic, where radiation doses have to be at least doubled to achieve the same effect (Suit and Lindberg 1968). The radiation reaction of skin made artificially hypoxic also demonstrates the radioprotectiveness of hypoxia (Dische et al. 1976).

Hyperbaric Oxygen and Sensitizers

There is also clinical evidence for hypoxic cells in human tumours. It should be recalled that Churchill-Davidson performed an experiment in eight patients where the effect of radiation under hyperbaric conditions was compared with that of radiation alone (Churchill-Davidson et al. 1955). The results were assessed histologically under blind conditions and in seven out of eight there was clearly a greater effect in hyperbaric oxygen; in the final case no tumour survived in either specimen. Included in the studies with misonidazole at Mount Vernon Hospital and The Royal Marsden Hospital, Surrey, were a total of 22 patients with multiple subcutaneous nodules (Thomlinson et al. 1976; Dawes et al. 1978; Ash et al. 1979). In 13 patients it was possible to make an assessment of tumour response and in nine of these an increased response was seen with misonidazole.

These clinical results can only be explained by the presence of hypoxic tumour cells which were sensitized by the drug.

A further important question is whether a system for sensitization of hypoxic tumour cells can lead to improved results when a comparison is made with the best fractionated course of radiotherapy. In this situation it must be recognized that the reoxygenation of hypoxic tumour cells occurs during a course of radiotherapy, so overcoming the problem of hypoxic cell resistance (Thomlinson 1969). It has yet to be shown in any tumour type that a method of hypoxic cell radiosensitization is so clearly superior to that of conventional radiotherapy that it should become the standard treatment.

Haemoglobin Level in Blood

If the oxygen concentration is important then anaemia should be a factor determining the effect of radiation. This indeed can be shown clinically. There is now a large body of evidence from animals and from man to relate anaemia and radiation effect. In animal studies the presence of anaemia can be shown to lead to a reduction in local tumour control by radiotherapy, and there are now ten clinical reports which give supporting evidence for this effect in the management of human tumours. Seven are concerned with the radiotherapy of carcinoma of the cervix (Evans and Bergsjø 1965; Hierlihy et al. 1969; Vigario et al. 1973; Shreiner et al. 1975; Bush et al. 1978; Overgaard 1985a; Balmukhanov et al. 1987). There is also evidence with regard to carcinoma of the bladder (Quilty and Duncan 1986), tumours in the head and neck region (Overgaard 1985b) and carcinoma of the bronchus (Dische et al. 1986a). There must be caution in interpretation of these data because there is a correlation between advancing disease and falling haemoglobin level, so there is a possibility that by separating out patients with a low haemoglobin those with an unfavourable prognosis are selected. The evidence, however, includes significant differences in local tumour control between high and low haemoglobin groups and the improvement in control which can be gained in anaemic patients by blood transfusion.

In the Mount Vernon Hospital series of patients with carcinoma of the bronchus, who were treated with opposing portals without cord shielding, there was a high incidence of radiation myelitis once the dose exceeded 33.5 Gy in the six fractions given over 17–18 days (Dische et al. 1981). In this group of patients at high risk for myelitis those who showed a haemoglobin level below 13 g% did not develop this complication; all the cases occurred in those with a higher haemoglobin level. There was also a direct correlation of haemoglobin level with survival (Dische et al. 1986a). There is, therefore, evidence that the radiation response of both tumour and normal tissue was related to haemoglobin concentration.

In the DAHANCA study of misonidazole in head and neck cancer it was shown that in the supra-glottic and pharyngeal carcinomas the haemoglobin concentration itself was directly associated with improved response (Overgaard 1985b). Of special interest was the fact that patients with a haemoglobin level in the lower range of normality (less than 14.5 g%) fared less well than those with a higher level. This evidence must influence the daily practice of radiotherapy and surely in attempting cure the haemoglobin level in all patients should be brought up to normal.

Lung Function

If anaemia is important then is lung function important? Impaired lung function so that the haemoglobin is not saturated with oxygen when it reaches the tumour might be expected to produce the same effect as though there were anaemia. In a preliminary review of the results of studies in 60 patients mainly with head and neck and bronchial carcinoma we have concluded that even though a considerable number of patients showed some impairment of lung function haemoglobin saturation levels were rarely below 90%. It would seem that at rest these patients were capable of fully saturating haemoglobin with oxygen. Therefore lung function is not likely to be a significant factor moderating radiation response except in those few patients in whom lung function is grossly impaired (Dische et al. unpublished data).

Relevance of Animal Models

As much of the data concerning the radiobiological effects of hypoxia comes from experimental work, is the animal tumour model an appropriate one for man? Hypoxic tumour cells are a significant cause of radioresistance in practically every tumour model studied. Adams has gathered a list of 44 of which 42 showed resistance due to hypoxia and, furthermore, such resistance was greatly reduced by the use of a chemical hypoxic cell sensitizer (G. E. Adams 1984, personal communication). There is, therefore, a remarkable contrast between the animal and the human situation.

Nearly all animal tumour models require the transplantation of tumour cells to a subcutaneous site, which may be in the tail, the foot or chest wall. In such a position the tumours can be observed, irradiated and excised with precision. But is this subcutaneous position a good model for the human tumour growing in its primary site?

Hill and Denekamp (1982) pursued a series of studies to determine the importance of site with regard to tumour growth and response to radiation, including radiation combined with hyperthermia or a hypoxic cell sensitizer. The pattern of results achieved was a complex one, but the outstanding feature is some of the differences encountered between tumours. The growth rates in three subcutaneous sites can be compared with that in an intramuscular site in the leg (Table 15.3). It would seem that in the very vascularized muscular

Table 15.3. Growth rates of sarcoma F tumour in CBA mouse according to site

Site of inoculation	Time to reach 5 mm diameter (days)	Volume doubling time (days)
Subcutaneous: chest wall	12	1.2
Subcutaneous: tail	33	1.7
Subcutaneous: foot	13	1.6
Intramuscular: leg	5	0.6

Modified from Hill and Denekamp (1982).

compartment of the leg, tumour growth occurs at a much more rapid rate than at any subcutaneous site; in the restricted subcutaneous tissues in the tail, where increase in size can be expected to lead to ischaemia, growth is slowest of all. This suggests that the vascular bed at the site of tumour is important in determining growth. Similar differences were seen with regard to the radiation responses, greater control being achieved with lower doses at the intermuscular site.

It seems probable, therefore, that the tumours which have been employed in animal experiments with radiosensitizers have been growing in poorly vascularized sites where the importance of hypoxia is likely to have been much greater than in a well-vascularized primary site such as is encountered in the head and neck region or in the lung. The subcutaneous tissues of the mouse may, however, be appropriate for modelling the growth of tumours in secondary sites such as the lymph nodes.

It is clear from the laboratory models that the radiation dose used in both single-dose and multiple-fraction studies is invariably higher than those employed in man. The dose to cure tumours 5–6 mm in diameter in only 50% of the animals is invariably beyond human tolerance even when multifraction regimes employing as many as 20 treatments are used. These differences may relate to higher concentrations of hypoxic cells in the animal models compared with the human primary tumour.

Tumour Perfusion

Further evidence that hypoxia influences the radiation response in man comes from data on vascular perfusion. Recent experiments in the laboratory in which drugs have been used to lower the blood pressure have led to massive necrosis of tumour (G. E. Adams 1986, personal communication). It appears that there is an inadequate control of blood flow in tumours compared with that in the normal tissues. As a response to hypotension the normal tissues adjust by vasodilation and in doing so substantially deprive the tumour of its circulation. In the converse situation, when the blood pressure rises the normal tissues again adjust and a greatly increased supply to the tumour can be expected.

We have performed analyses on our patients with carcinoma of the bronchus in whom there was radiation myelitis above a threshold dose of 33.5 Gy and where the haemoglobin concentration was significantly related both to radiation myelitis and to survival (Dische et al. 1986a). Here, as a quite independent variable, patients with a higher systolic blood pressure or larger pulse pressure survived longer than those with a lower one. The incidence of radiation myelitis was unaffected (Dische et al. unpublished data). This clinical evidence is entirely compatible with that now coming from the laboratory concerning vascular perfusion. These observations give further support to the importance of oxygen availability at the time of radiotherapy.

Histological Evidence of Hypoxic Cells in Human Tumours

Urtasun and his colleagues at Edmonton have administered radiolabelled misonidazole to patients before removal of tumours (Urtasun et al. 1986). They demonstrated a significant radiolabelling of cells presumed to be hypoxic in only 5 of 11 tumours studied. The tumours with hypoxic cells were mainly epithelial ones while those showing no detectable concentration of misonidazole were mainly sarcomas. Caution is needed in the interpretation of these results as the fixation of misonidazole adducts in tissues demands the presence of the sensitizing agent in hypoxic tissue for a number of hours and this may not correspond to hypoxia at the time of radiotherapy. These studies are nevertheless important, and more are needed.

Relevance of Tumour Cell Kinetics to the Concept of Hypoxia

The evidence that human tumours have a potential for rapid growth has accumulated during recent years. The techniques recently developed in the Gray Laboratory have enabled us to learn the labelling index and the cell cycle time of human tumours and from them the potential doubling time (Wilson et al. 1985). The technique requires the intravenous administration of bromodeoxyuridine several hours before sampling, but within 24 hours these important tumour cell kinetic data are available. In 10 of 14 patients studied the potential doubling time of the tumour was found to be less than 6 days (Wilson et al. unpublished data). It is in such rapidly growing tumours that a high and significant concentration of hypoxic cells would be expected.

Accelerated courses of radiotherapy have been most promising in these cases. In our series use of a 36-fraction regime given on 12 consecutive days produced complete regression in the primary site in all except one of 20 patients with advanced head and neck tumours (Saunders et al. 1987).

It can be argued that with rapidly growing tumours the main cause for radiation failure in a well-vascularized site is likely to be the repopulation between fractions and not hypoxia. It is of interest, however, that the results observed in the secondary nodes of these patients have been less satisfactory and perhaps it is here in this less well vascularized area that hypoxia is an additional major problem. It is probable that to achieve the best control with very rapidly growing tumours hyperfractionation must be combined with selective hypoxic cell sensitization.

Evidence that Hypoxic Tumour Cells are Responsible for Failure, Despite the Best Fractionated Radiotherapy

A re-analysis of patients treated for carcinoma of the cervix using hyperbaric oxygen in three radiotherapy centres showed a remarkable finding in the subgroup of patients who were so severely anaemic prior to presentation that blood transfusion was required before radiotherapy. When treated in hyperbaric oxygen there was a high local tumour control rate while conventional radiotherapy resulted in particularly poor results (Dische et al. 1983). This appeared to identify one special subgroup where hypoxia was a severe problem which could not be overcome by routine fractionated radiotherapy; however a radiosensitizing method was successful.

In two of the studies of misonidazole in head and neck cancer significant and sustained benefit was achieved, giving further examples of tumours where, despite fractionated radiotherapy, hypoxia remains a problem and a sensitizer can help (Sealy and Cridland 1984; Overgaard 1985b).

Further Techniques for the Exploitation of the Oxygen Effect

Improved Oxygen Transport

In addition to the use of blood transfusion, hyperbaric oxygen and chemical sensitizing agents, improved oxygenation of a tumour may be achieved in other ways. Perfluorochemical agents improve oxygen transport and when used with concurrent 100% oxygen-breathing have produced promising results recently in a pilot study in head and neck cancer (Rose et al. 1986). The oxygen requirements of tissue can be reduced by hypothermia leading to a rise in oxygen availability, and this has been combined with radiotherapy in hyperbaric oxygen in a promising pilot study (Sealy et al. 1986). Improved vascular perfusion of tumours may be achieved by manipulation of the blood pressure and this is a new area for clinical experiment.

Radiotherapy under Hypoxia

An alternative approach is the deliberate use of hypoxic conditions during radiotherapy. In the past a number of groups have rendered limbs hypoxic by temporary arrest of the circulation, thus reducing the dominance of hypoxia in tumour over that in normal tissue (Suit and Lindberg 1968). Results did not, however, suggest an advantage over conventional radiotherapy. Perhaps this was because the technique required the use of unsatisfactory fractionation. Centres in East Germany and in the Soviet Union are currently testing hypoxic radiotherapy with patients breathing 10% oxygen in the period before and during treatment (Tysb et al. 1983). Such a technique should lead to protection of normal tissues. In contrast, evidence accumulating in the laboratory suggests that the blood supply to many tumours is, under normal conditions, barely adequate and that a further decrease or a reduction in oxygen supply can induce severe hypoxia and necrosis (G. E. Adams 1986, personal communication). It remains to be seen whether, on balance, the therapeutic ratio will be in favour of greater effects in tumour compared with those in normal tissues and whether the technique can be reliably employed. A similar effect may also be achievable with the use of drugs which shift the oxygen–haemoglobin dissociation curve to the left so rendering it less efficient and giving the effect of a temporary anaemia.

Hypoxia before Radiotherapy

Another approach is to stimulate the conditions in the patients with carcinoma of the cervix who had severe anaemia and give a pre-treatment period of hypoxia prior to radiotherapy under well-oxygenated conditions. R. Sealy (personal communication 1986) has deliberately induced anaemia in patients and then after restoration of the haemoglobin concentration treated them in hyperbaric oxygen. Drugs which move the haemoglobin dissociation curve to the left could similarly be used, as also could methods to reduce blood flow to tumour.

New Chemical Radiosensitizers

The main line of research remains with the use of chemical agents to sensitize hypoxic tumour cells. Two drugs, SR 2508 and Ro 03–8799, have now completed phase I and phase II testing (Coleman et al. 1984; Saunders et al. 1984). In the former the dose-limiting effect is again peripheral neuropathy, but considerably higher doses may be given than of misonidazole. With Ro 03–8799 peripheral neuropathy does not occur, but there is an immediate disturbance giving rise to the complaint of heat and malaise which limits the amount which can be given on any one occasion; the drug does show a several-fold concentration in tumours however (Saunders et al. 1984).

Tumour concentration studies performed when two or three drugs are given simultaneously enable an estimate of the radiosensitizing concentrations likely to be achieved and allow a direct comparison with those obtainable with misonidazole in the same human tumour (Dische et al. 1986b). The conclusions from these studies are that in multifraction radiotherapy both are likely to show concentrations which will give a sensitization equal to a fivefold increase in misonidazole dosage, that is as though a cumulative dose of 60 g instead of 12 g/m^2 of surface area had been given.

The two sensitizing drugs have different toxic effects, so their use in combination could give greater sensitization combined with equal or reduced morbidity. A phase I dose escalation study has been completed (Newman et al. 1986). Unfortunately problems of drug availability and anticipated problems with drug regulation authorities have considerably hampered this development.

Particularly high concentrations of Ro 03–8799 have been found in malignant melanoma which may be tenfold greater than in the plasma at the time of tumour sampling, and promising results have been found in a pilot study combining Ro 03–8799 with radiotherapy. Five of seven patients treated with the objective of cure have shown complete and lasting local clearance (Dische unpublished data). Randomized controlled clinical trials have commenced on an international basis on the use of SR 2508 in head and neck cancer and Ro 03–8799 in advanced carcinoma of the cervix.

Conclusion

The oxygen effect remains a radiobiological phenomenon of the utmost importance. There is great potential for improving the results of cancer treatment, but such advance is dependent on improved biological measurement of human tumours and the evolution of integrated plans for radiotherapy.

References

Adams GE, Dewey DL (1963) Hydrated electrons and radiobiological sensitization. Biochem Biophys Res Commun 12:473–477

Ash D, Peckham M, Steel JJ (1979) The quantitative response of human tumours to radiation and misonidazole. Br J Cancer 40:883–889

Asquith JC, Watts ME, Patel K, Smithen CE, Adams GE (1974) Electron-affinic sensitization. 5. Radiosensitization of hypoxic bacteria and mammalian cells in vitro by some nitroimidazoles and nitropyrazoles. Radiat Res 60:108–118

Balmukhanov S, Rismohamedova R, Aitkoolova Z, Revész L (1987) Radiosensization of cervix cancer in anemic patients by metronidazole treatment. Br J Radiol (in press)

Bleehen NM (1983) Heat and drugs: current status of thermo-chemotherapy. In: Steel GG, Adams GE, Peckham MJ (eds) The biological basis of radiotherapy. Elsevier, Amsterdam, pp 321–332

Bush RS, Jenkin RDT, Allt WE et al. (1978) Definitive evidence for hypoxic cells influencing cure in cancer therapy. Br J Cancer 37 [Suppl III]:302–306

Churchill-Davidson I, Sanger D, Thomlinson RH (1955) High pressure oxygen and radiotherapy. Lancet I:1091–1095

Coleman CN, Urtasun RC, Wasserman TH et al. (1984) Initial report of the phase I trial of the hypoxic cell radiosensitizer SR 2508. Int J Radiat Oncol Biol Phys 10:1749–1753

Dawes PJDK, Peckham MJ, Steel GG (1978) The response of human tumour metastases to radiation and misonidazole. Br J Cancer 37 [Suppl III]:290–296

Dische S, Gray AJ, Zanelli GD (1976) Clinical testing of the radiosensitizer Ro 07–0582. 2. Radiosensitization of normal and hypoxic skin. Clin Radiol 27:159–166

Dische S, Saunders MI, Lee ME, Adams GE, Flockhart IR (1977) Clinical testing of the radiosensitizer Ro 07–0582. Experience with multiple doses. Br J Cancer 35:567–579

Dische S, Martin WMC, Anderson P (1981) Radiation myelopathy in patients treated for carcinoma of bronchus using a six fraction regime of radiotherapy. Br J Radiol 54:29–35

Dische S, Anderson PJ, Sealy R, Watson ER (1983) Carcinoma of the cervix – anaemia, radiotherapy and hyperbaric oxygen. Br J Radiol 56:251–256

Dische S, Saunders MI, Warburton MF (1986a) Hemoglobin, radiation, morbidity and survival. Int J Radiat Oncol Biol Phys 12:1335–1337

Dische S, Saunders MI, Dunphy EP et al. (1986b) Concentrations achieved in human tumors after administration of misonidazole, SR–2508 and Ro 03–8799. Int J Radiat Oncol Biol Phys 12:1109–1111

Evans JC, Bergsjø P (1965) The influence of anaemia on the results of radiotherapy in carcinoma of the cervix. Radiology 84:709–717

Foster JL, Willson RL (1973) Radiosensitisation of anoxic cells by metronidazole. Br J Radiol 46: 234–235

Gray AJ, Dische S, Adams GE, Flockhart IR, Foster JL (1976) Clinical testing of the radiosensitizer Ro 07–0582. 1. Dose tolerance, serum and tumour concentration. Clin Radiol 27:151–157

Gray LH, Conger AO, Ebert M, Flockhart IR, Foster JL (1953) The concentration of oxygen dissolved in tissue at the time of irradiation as a factor of radiotherapy. Br J Radiol 26:638–648

Hierlihy P, Jenkin RDT, Stryker JA (1969) Anemia as a prognostic factor in cancer of the cervix. Can Med Assoc J 100:1100–1102

Hill SA, Denekamp J (1982) Site dependent response of tumours to combined heat and radiation. Br J Radiol 55:905–912

Mottram JC (1935) Annual Report of the Mount Vernon Hospital and Radium Institute, Northwood, England.

Newman HFV, Bleehen NM, Workman P (1986) A phase I study of the combination of two hypoxic cell radiosensitisers, Ro 03–8799 and SR–2508: toxicity and pharmacokinetics. Int J Radiat Oncol Biol Phys 12:1113–1116

Overgaard J (1985a) Misonidazole combined with radiotherapy in the treatment of carcinoma of the uterine cervix. In: Proceedings of the conference on chemical modifiers of cancer treatment, Clearwater, Florida, 20–24 October 1985

Overgaard J (1985b) Misonidazole combined with split-course radiotherapy in the treatment of invasive carcinoma of larynx and pharynx. In: Proceedings of the conference on chemical modifiers of cancer treatment, Clearwater, Florida, 20–24 October 1985

Quilty PM, Duncan W (1986) The influence of hemoglobin level on the regression and long term local control of transitional cell carcinoma of the bladder following photon irradiation. Int J Radiat Oncol Biol Phys 12:1735–1742

Rose C, Lustig R, McIntosh N, Teicher B (1986) A clinical trial of fluosol DA 20% in advanced squamous cell carcinoma of the head and neck. Int J Radiat Oncol Biol Phys 12:1325–1328

Saunders MI, Dische S, Anderson PJ, Tothill M, Stratford MRL, Minchinton AI (1984) The clinical testing of Ro 03–8799: pharmacokinetics, toxicology, tissue and tumour concentrations. Int J Radiat Oncol Biol Phys 10:1759–1763

Saunders MI, Dische S, Fowler JF et al (1987) Radiotherapy employing 3 fractions on each of 12 consecutive days. Acta Radiol (in press)

Sealy R, Cridland S (1984) The treatment of locally advanced head and neck cancer with misonidazole, hyperbaric oxygen and irradiation: an interim report. Int J Radiat Oncol Biol Phys 10:1721–1723

Sealy R, Harrison GG, Morrell D et al. (1986) A feasibility study of a new approach to clinical radiosensitisation: hypothermia and hyperbaric oxygen in combination with pharmacological vasodilation. Br J Radiol 59:1093–1098

Shreiner P, Siracka E, Siracky J, Manka I (1975) The effect of anemia on the radiotherapy results of the uterine cervix cancer. Neoplasma 22:655–660

Steel GG (1983) The combination of radiotherapy and chemotherapy. In: Steel GG, Adams GE, Peckham MJ (eds) The biological basis of radiotherapy. Elsevier, Amsterdam, pp 239–248

Suit H, Lindberg R (1968) Radiation therapy administered under conditions of tourniquet-induced local tissue hypoxia. Am J Roentgenol 102:27–37

Thomlinson RH (1969) In: Proceedings of Carmel conference on time and dose relationships in radiation biology as applied in radiotherapy. Brookhaven National Laboratory report no. 50203, p 242

Thomlinson RH, Dische S, Gray AJ, Errington LM (1976) Clinical testing of the radiosensitizer Ro 07–0582. 3. Response of tumours. Clin Radiol 27:167–174

Tysb AF, Konoplayannikov AG, Ocvhnnikova VG (1983) The importance of the modifiers in radiotherapy of malignant neoplasms. Radiosensitization Newsletter 2:7–8

Urtasun RC, Chapman JD, Raleigh JA, Franko AJ, Koch CJ (1986) Binding of ^{3}H-misonidazole to solid human tumors as a measure of tumor hypoxia. Int J Radiat Oncol Biol Phys 12:1263–1267

Vigario G, Kurohara SS, George FW (1973) Association of hemoglobin levels before and during radiotherapy with prognosis in uterine cervix cancer. Radiology 106:649–652

Wilson GD, McNally NJ, Dunphy E, Kärcher H, Pfragner R (1985) The labelling index of human and mouse tumours assessed by bromodeoxyuridine in vitro and in vivo and flow cytometry. Cytometry 6:641–647

16 Combined Treatment with Radiation and Anti-Cancer Drugs: Experimental and Clinical Results

H. Bartelink, A. C. Begg, L. Dewit and F. A. Stewart

Introduction

The combination of radiotherapy and chemotherapy for the treatment of patients with malignant tumours has found increasing use during the last decade. The main aims of this combined treatment approach are to increase the control of primary tumours and to eradicate microscopic metastases. The first of these aims, improvement of local control, can be attempted in two ways: (*a*) giving the chemotherapy separated in time from the irradiation, or (*b*) administering cytotoxic drugs during the irradiation period in order to obtain a radiosensitizing effect. With the first approach only additive cell killing is sought, while using cytotoxic agents as radiosensitizers may cause supra-additive cell killing by specific interactions between the two agents. In either situation the chemotherapy will also have an effect on microscopic metastases. Additive or supra-additive results of a combined treatment can, however, also occur in normal tissues. The anticipated therapeutic gain will then be decreased due to additional complications in normal tissue or even the induction of second malignancies. It is therefore important that, before introducing promising new combinations of radiotherapy and chemotherapy, studies in animals should be carried out to quantitate the effect of the combination on normal tissues as well as on tumours. We will describe experimental results for combined treatments with cisplatin and X-rays, concentrating mainly on results obtained in our own laboratory, in order to illustrate some general principles of combined modality therapy. This will be followed by an evaluation of clinical trials in which the combination of irradiation and chemotherapy has been investigated.

Experimental Tumour Studies

In the following section the experimental approach to investigating interactions between chemotherapy and radiotherapy will be illustrated by our studies with cisplatin. The aim of these studies was to define the relative roles of additive killing, radiosensitization (modification of the single dose radiation survival curve), inhibition of repair (sublethal and potentially lethal damage) and inhibition of proliferation.

Killing and Radiosensitization

Fig. 16.1 shows the magnitude of cell killing and radiosensitization by a series of cisplatin doses on log-phase, aerobic RIF1 cells in culture (Begg et al. 1986). The left-hand panel shows, firstly, that increasing drug doses caused progressive cytotoxicity in the absence of radiation (*y* intercept values). Secondly, cells irradiated at the end of the 1-hour drug treatment were more sensitive to X-rays than cells not previously exposed to cisplatin. This is more clearly seen in the right-hand panel, where the curves have been corrected for drug-induced cell killing. Sensitizer enhancement ratios (SERs) can be calculated from these corrected survival curves by reading off the doses with and without drug to produce a given level of killing. SERs of over 2 were obtained with cisplatin

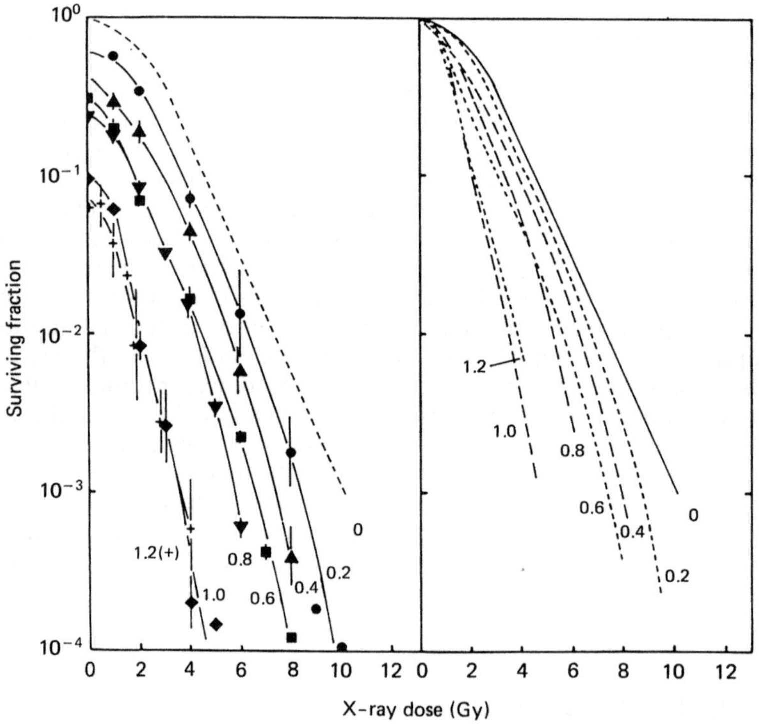

Fig. 16.1. Radiosensitization of RIF1 cells by cisplatin. Cells were irradiated at the end of a 1-hour exposure while still in contact with the drug. Numbers against the curves are cisplatin concentrations in μg/ml. The curves in **b** have been reproduced from those in **a** after correcting for cell killing by drug alone.

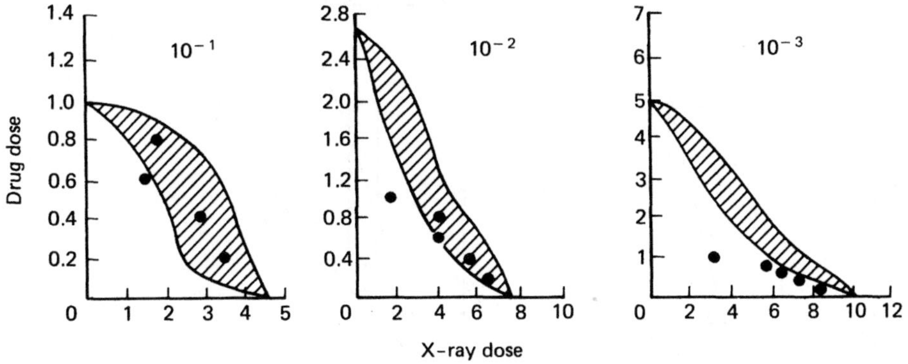

Fig. 16.2. Isobologram analysis of the interaction between cisplatin and X-rays. The *hatched areas* are the "envelopes of additivity" calculated from the single agent survival curves. The points are the combined treatment data and show the doses of X-rays (Gy) and drug (μg/ml; 1 hour) required to produce the surviving fraction indicated in each panel.

concentrations which are achievable in the blood of man, indicating that this drug can be a potent radiosensitizer.

Further analysis can be performed with these data in order to test whether the observed interactions are more than additive. Construction of isobolograms is one such analysis that is particularly useful in cases where one or both dose–response curves for the individual agents to be considered are non-linear (Steel and Peckham 1979). This is the case for both cisplatin and radiation in RIF1 cells (Begg et al. 1986), for which the resultant isobolograms are shown in Fig. 16.2. The position of the combined treatment data points relative to the "additivity envelopes" depends on the level of effect taken for the analysis, but at low survival levels the points lie beneath the envelope, indicating a supra-additive interaction. Radiosensitization by cisplatin of several different tumour types in vivo has also been observed. In addition to SERs of over 2 in subcutaneous or intramuscular RIF1 tumours (Figs. 16.3 and 16.4), marked enhancements have also been seen in a C3H mammary tumour (Overgaard and Khan 1981), the MTG-B mammary tumour (Douple and Richmond 1982) and in the SCC VII hypoxic tumour cells after systemic administration of well-tolerated doses.

Oxic Versus Hypoxic Radiosensitization

Cisplatin has previously been shown to be a radiosensitizer of hypoxic cells (Douple and Richmond 1980). The data presented above and those of others, however, demonstrate that its radiosensitizing capacity is not limited to hypoxic cells. Ziegler (1986), for example, studied radiosensitization by cisplatin in four cell lines – two rodent and two human. In only one of the lines was the drug a specific sensitizer of hypoxic cells, whereas in two lines sensitization was seen only in oxic cells. In the fourth cell line no radiosensitization was observed. The

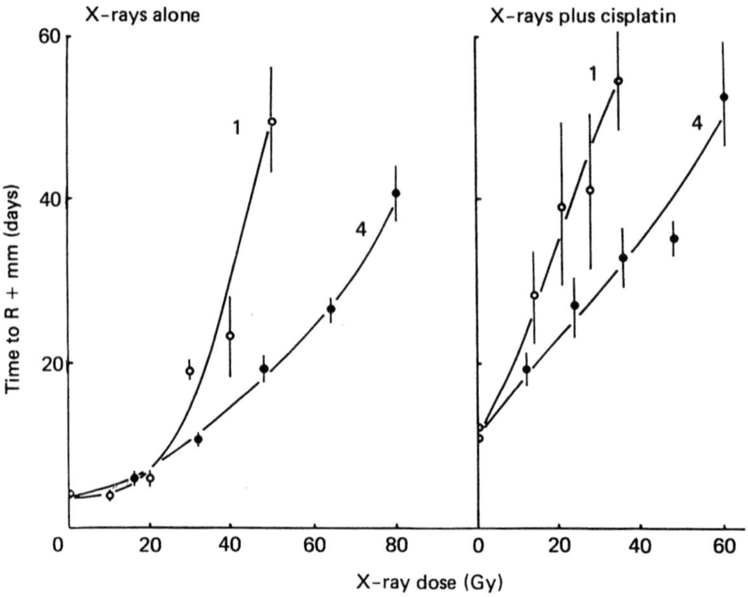

Fig. 16.3. Effect of cisplatin on split dose repair. Hypoxic (clamped) RIF1 tumours were given one fraction of X-rays (*open circles*) or four fractions (*filled circles*) separated by 5-hour intervals, with (*right*) or without (*left*) a single dose of cisplatin (6 mg/kg) given 30 minutes before the first irradiation. The end-point was the time taken for tumours to grow to 2 mm above the size at treatment. Similar separations between the 1 and 4 fraction curves on the two graphs indicate little or no drug-induced inhibition of repair.

reason why sensitization is seen in some lines but not others is not known. In the relatively few studies completed so far there does, however, appear to be a relationship between sensitivity to killing by the drug and radiosensitization (Begg et al. 1986). RIF1 cells are one of the cell lines most sensitive to cisplatin and show a large degree of radiosensitization. CHO cells are more than 10 times as resistant to the drug and show only a small degree of sensitization (Murthy et al. 1979). Studies are urgently needed to elucidate mechanisms which may then lead to ways of predicting which cells or tumours would benefit from treatment with cisplatin as a radiosensitizer.

Effects on Repair of Radiation Damage

In addition to its ability to modify the shape of the radiation cell survival curve, cisplatin has also been shown to inhibit repair of both sublethal and potentially lethal radiation damage (SLD and PLD) (Dritschilo et al. 1979; Carde and Laval 1981). Repair inhibition can be almost complete but is dependent on the time of administration of the drug relative to irradiation, as illustrated for RIF1 cells in Fig. 16.5. Further studies on RIF1 cells, however, showed that cisplatin appeared to have opposite effects on the two types of repair in this cell line. Inhibition of SLD repair (increase in survival between two radiation doses) was

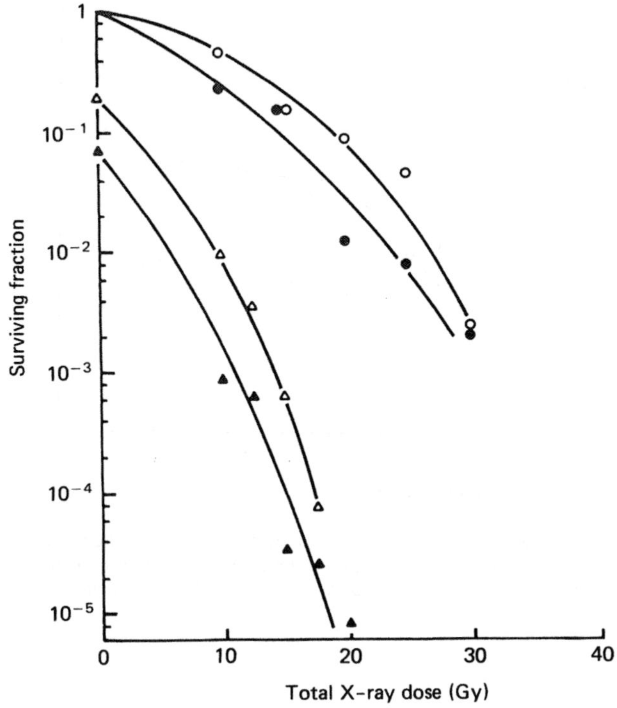

Fig. 16.4. Radiosensitization by cisplatin of intramuscular RIF1 tumours given five fractions of X-rays (4 Gy) alone (*circles*) or X-rays (4 Gy) plus drug (1.6 mg/kg) (*triangles*). Cell survival was assayed in vitro immediately (●,▲) or 24 hours (○,△) after each dose. Both cell killing and radiosensitization are evident in the combined treatment groups (△,▲) with no evidence of inhibition of potentially lethal damage repair.

confirmed, whereas PLD repair (increase in survival after a single dose) was significantly enhanced by cisplatin (Table 16.1). The net result of the two opposing effects was a small but insignificant decrease in total repair. The contrast between the stimulation of PLD repair by cisplatin observed in RIF1 cells and its inhibition in CHO cells (Dritschilo et al. 1979) and rat hepatoma cells (Carde and Laval 1981) shows that this effect, like many others, is cell line dependent, and indicates the important need for a more basic understanding of PLD repair at the molecular level.

The effects of cisplatin on radiosensitization and repair in RIF1 tumours treated in vivo were also investigated. Fig. 16.4 shows the marked radiosensitization obtained by the drug in a five-fraction treatment schedule in which cell survival was assayed in vitro either immediately or 24 hours after each dose (Bartelink et al. 1986). These data show no PLD repair inhibition by cisplatin. Fig. 16.3 shows the radiosensitizing effects of cisplatin, again in RIF1 tumours, in both single dose and four-fraction schedules as assayed by growth delay. From this figure one can also assess the effect of the drug on split dose repair, from the relative displacements of the 1 and 4 fraction curves with and without drug. This analysis showed, in agreement with the in vitro findings, that at equivalent total

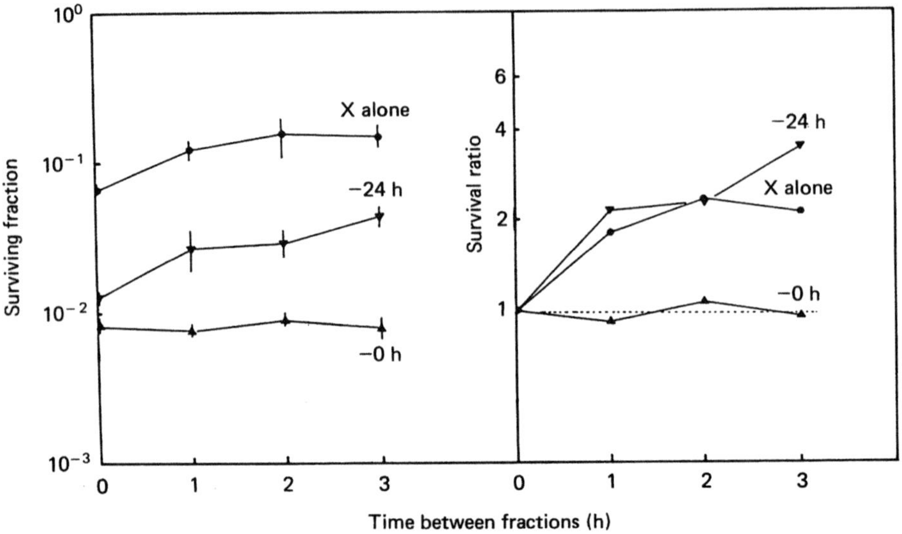

Fig. 16.5. Inhibition of SLD repair by cisplatin in RIF1 tumour cells, as determined by the increase in survival when 5 Gy was split into two doses of 2.5 Gy. A 1-hour exposure of 1 μg/ml cisplatin was given, ending immediately (−0 h) or 24 hours (−24 h) before the first irradiation. The *right-hand panel* shows survival as a ratio of the 0 h (single dose) value.

radiation doses there was little or no effect on repair of radiation damage. Despite this lack of repair inhibition, considerable enhancement of radiation damage was seen for both schedules. There have been very few other tumour studies specifically investigating repair inhibition by cisplatin in vivo, although Fu and colleagues (1985) reported that cisplatine produced a decrease in the split dose recovery ratio in a murine squamous cell carcinoma.

Table 16.1. Cisplatin: effects on X-ray repair in vitro

Survival ratio[a]						
X-rays alone				X-rays plus cisplatin		
Group[b]	PLDR[c]	SLDR[d]	Total	PLDR	SLDR	Total
1 L	0.93	2.84	2.31	1.28	1.18	1.50
2 L	1.21	1.96	2.38	1.32	1.49	1.98
4 L	1.07	1.54	1.65	1.25	1.07	1.34
4 P	1.16	1.51	1.75	2.08	1.13	2.34
Mean	1.09	1.87	2.02	1.48	1.22	1.79
±SEM	±0.06	±0.20	±0.19	±0.20	±0.09	±0.23

A. C. Begg, J. Emondt and H. Bartelink (unpublished data).
[a] Factor increase in surviving fraction due to repair.
[b] Days between plating and irradiation. L, log phase; P, plateau phase.
[c] PLDR, potentially lethal damage repair.
[d] SLDR, sublethal damage repair.

Effects on Proliferation

Another potentially important mechanism whereby a drug can increase the effects of a fractionated irradiation schedule is by the inhibition of proliferation. Consequently, this was investigated on the sensitive RIF1 tumour cells in culture. It was found that despite either one or two logs of cell killing, the survivors of the cisplatin treatment proliferated at the same rate as untreated cells, with little or no evidence of an initial delay (Fig. 16.6). This was also true after irradiation with doses which gave the same degree of cell killing as the drug treatments (the RIF1 cell line has a "normal" radiosensitivity; $D_0 = 1.2$ Gy).

Experiments have also been carried out on RIF1 cells during the first 24 hours after cisplatin treatment using a monoclonal antibody against bromodeoxyuridine incorporated into DNA, to investigate in detail the effect of the drug on cell cycle kinetics. The results confirmed those of the growth studies described

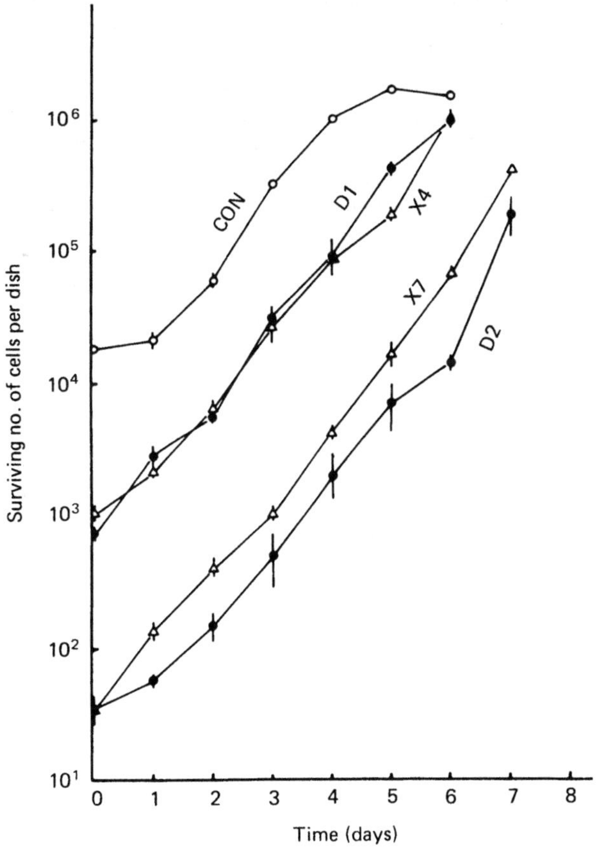

Fig. 16.6. Effect of cisplatin on proliferation of RIF1 tumour cells in culture. Numbers of cells surviving drug or X-ray treatments were assessed as a function of time by the colony formation assay. D1 and D2, 1 and 2 μg/ml cisplatin; X4 and X7, 4 and 7 Gy X-rays; CON, control cells. Parallel slopes indicate little or no inhibition of proliferation.

above, in showing little or no effect of the drug on cell cycle progression, despite considerable cell killing. Inhibition of proliferation does not, therefore, appear to be an important mechanism of action of cisplatin in these drug-sensitive RIF1 cells. Cisplatin has been shown in some cell lines to inhibit DNA synthesis (Harder and Rosenberg 1970) and to cause dose-dependent cell cycle perturbations (Bergerat et al. 1979). Distinctions between doomed cells and surviving cells were not made, however.

The conclusions from these in vitro and in vivo tumour studies are that radiosensitization (i.e. modification of the radiation cell survival curve), plus additive cell killing, are the two most important mechanisms by which cisplatin can increase the effects of radiotherapy. The drug is capable of inhibiting repair of radiation damage, but in vivo studies suggested that this is not a major factor. Inhibition of proliferation also does not appear to be a major effect. Studies are now under way to investigate which molecular lesion, such as cisplatin–DNA adducts, is the most important for radiosensitization and/or repair inhibition. These studies employ cell lines with different sensitivities to the cytotoxic and radiosensitizing actions of cisplatin and a range of drug–DNA adducts are being measured. Included in this work are antibodies and an antiserum which recognize specific cisplatin–DNA adducts (Poirier et al. 1982; Terheggen et al. 1987). From such studies it is hoped to gain insight into the mechanism of radiosensitization by cisplatin and, ultimately, to be able to predict which tumours will be radiosensitized by the drug and, just as importantly, which will not. Other mechanisms by which cisplatin may radiosensitize cells include reduction of intracellular endogenous protectors such as glutathione, and fixation of damage by virtue of the electron affinity of the drug, analogous to the mechanism of action of hypoxic cell radiosensitizers such as misonidazole (Douple and Richmond 1980). The precise role of these two factors in cell lines with differing SERs for cisplatin is not known and merits further study.

Experimental Normal Tissue Studies

Any increase in normal tissue reaction which occurs as a result of combined chemotherapy and radiotherapy, whether due to additive toxicities or interaction, must be taken into account when assessing the possible therapeutic benefit (Phillips et al. 1975). To illustrate such effects, studies on the response of a range of normal tissues in mice to combined treatments of X-rays and cisplatin, using both single dose and fractionated schedules, will be described. The systems investigated were acute damage in the small intestine (duodenum) and late damage in the large intestine (rectum) and in the kidney.

Small Intestine

Duodenal damage was measured using the microcolony assay of Withers and Elkind (1970). Fig. 16.7 illustrates the crypt cell survival after single doses of X-rays alone or in combination with 10 mg/kg cisplatin given 30 minutes before or 10 minutes after irradiation. Cisplatin caused a downward shift in the X-ray

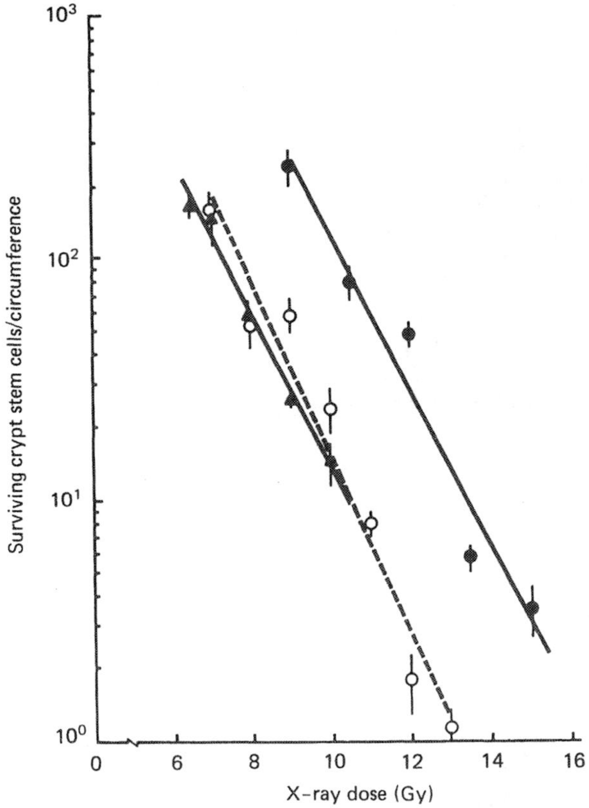

Fig. 16.7. Duodenal crypt stem cell survival versus X-ray dose for single doses of X-rays alone (●) or in combination with 10 mg/kg of cisplatin given 30 minutes before (▲) or 10 minutes after X-rays (○). The curves are obtained by linear regression analysis. Error bars are ±1 SEM.

survival curve with a dose enhancement factor (DEF)* of 1.3. There was, however, no modification of the slope of the survival curve, which suggests that there was no radiosensitization by the drug but only independent killing by drug and radiation. Separate experiments demonstrated that cisplatin given at various intervals of up to 12 hours before or 6 hours after irradiation resulted in similar levels of crypt cell killing (Dewit et al. 1985a), supporting the hypothesis of independent cell killing by drug and X-rays in the duodenum. These results are in agreement with those of Luk et al. (1979), Phillips (1979) and Shenken et al. (1979), but contrast with those of Von der Maase (1984a) who reported radiosensitization in duodenal crypts by cisplatin which was only observed when the drug was given before irradiation.

To assess the influence of cisplatin on SLD repair in the intestine, a series of fractionated irradiations were given and repair after X-rays alone was compared

---

$$\text{* DEF} = \frac{\text{Dose of X-rays alone}}{\text{Dose of X-rays + drug}}$$ for an equivalent level of cell killing or tissue damage.

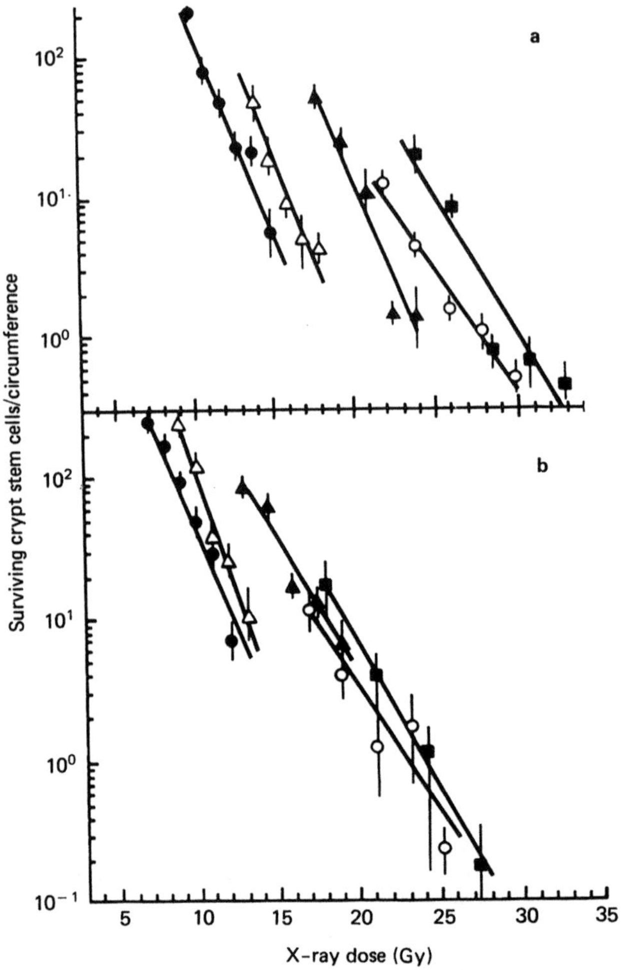

Fig. 16.8. Duodenal crypt stem cell survival as a function of X-ray dose for X-rays alone (**a**) and cisplatin plus X-rays (**b**) after single doses (●) and 2 (△), 5 (▲), 10 (○) or 15 fractions (■). The curves were obtained by weighted linear regression analysis. Vertical bars are ±1 SEM.

with that seen after X-rays plus drug (Dewit et al. 1985b).

For these experiments a single dose of 8 mg/kg was injected before the first of 1, 2, 5, 10 or 15 X-ray doses given at 3-hour intervals. After X-rays alone the total dose for a constant level of cell survival increased with increasing number of X-ray fractions due to SLD repair (Fig. 16.8a). After the combined treatment there was also an increase in total dose with fractionation (Fig. 16.8b), but all the curves were shifted to lower X-ray doses. The extent of the vertical shift of the survival curves for X-rays plus drug compared with X-rays alone was essentially independent of the fractionation scheme (Fig. 16.9) and was similar to the calculated value (from separate experiments) of approximately 1 $\log_{10}$ of cell killing by 8 mg/kg cisplatin alone. These data suggest that independent cell

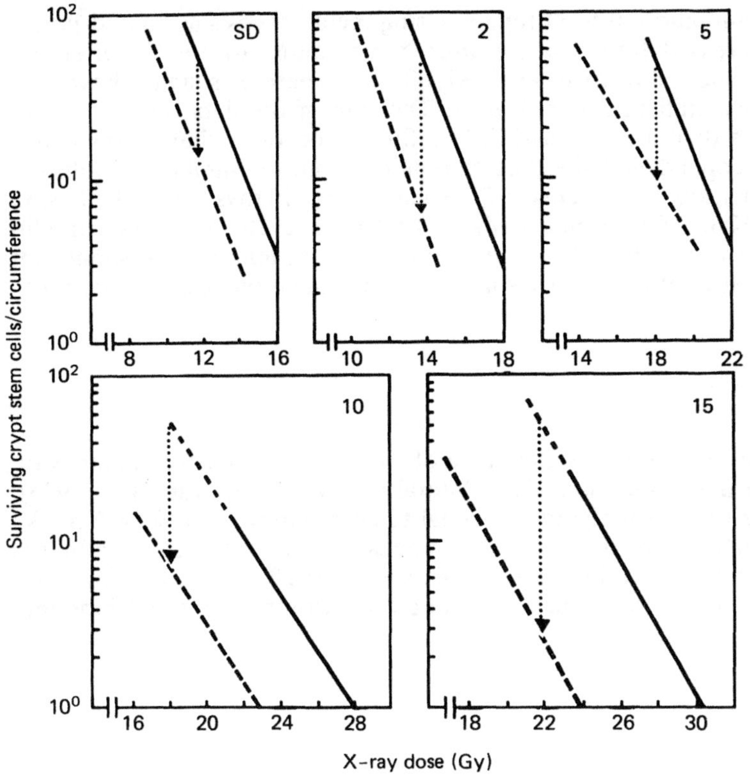

Fig. 16.9. Comparison of the crypt stem cell survival curves versus dose after X-rays (*continuous lines*) and cisplatin plus X-rays (*dashed lines*), and their vertical distance (*dotted lines*) within each fractionation schedule. SD, single dose; 2,5, etc., the number of fractions.

killing dominates the response to combined treatments. Linear-quadratic (LQ) analyses of the fractionated data, however, demonstrated a slight increase in the α/β ratio (an indication of X-ray survival curve shape, inversely related to repair capacity) from 13.0 ± 1.7 Gy for X-rays alone to 20 ± 4 Gy for the combined treatments. This would be consistent with there being a small amount of inhibition of SLD repair by cisplatin in the duodenum, as has been observed in experiments by Luk et al. (1979). The contribution of this effect is small, however, compared with independent cell killing.

Large Intestine

The influence of combined treatments of cisplatin and X-rays on the development of late rectal damage was also studied, using diarrhoea and rectal stenosis as end-points (Dewit et al. 1987). There was no increase in the incidence of diarrhoea or the development of rectal stenosis after the combined treatment compared with X-rays alone for drug given before or after single dose irradiation. When cisplatin was combined with fractionated irradiation there was a trend towards inhibition of SLD repair. Using rectal stenosis as an end-point, the α/β ratio increased from 4.4 Gy after X-rays alone to 6.9 Gy after the combination treatment (Dewit et al. 1987). This increase was small, however, and failed to reach statistical significance. The sum of the data on the rectum therefore suggests that cisplatin had little influence on the radiation response, either through independent cell killing or repair inhibition. Similarly, published data for combination treatments of cisplatin and X-rays in rodent skin (Overgaard and Kahn 1981; Von der Maase 1984b; Lelieveld et al. 1985) and lip mucosa (Landuyt et al. 1986) have generally failed to demonstrate significant drug toxicity or any influence of cisplatin on the radiation response of these tissues.

Kidney

Renal damage was measured by clearance of ^{51}Cr-EDTA after treatments with cisplatin alone or in combination with bilateral irradiation (Stewart et al. 1986). After drug alone there was significant renal toxicity within 1 week of 6 mg/kg cisplatin, with partial recovery from 1 to 7 weeks (Fig. 16.10). After irradiation alone there was no measurable damage before about 20 weeks, but from this time there was a progressive and dose-related deterioration in renal function.

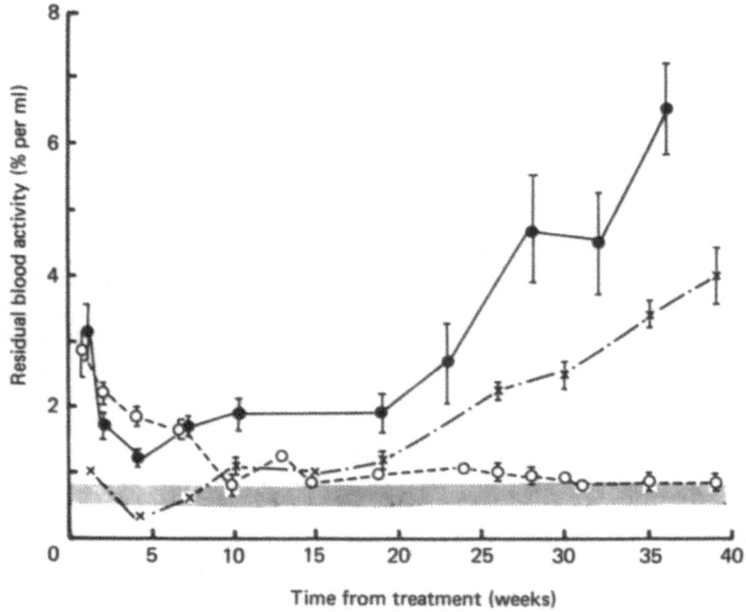

Fig. 16.10. Time course for changes in renal function (assessed from clearance of ^{51}Cr-EDTA at 30 minutes following injection) after cisplatin, X-rays or a combination of the two treatments. Each point represents the mean (±1 SEM) of a group of 4–10 mice. The response of untreated, control mice is indicated by the *stippled area*. ○, 6 mg/kg cisplatin alone; ×, 14 Gy X-rays alone; ●, 6 mg/kg cisplatin ½ h before 14 Gy X-rays.

Cisplatin given before or after X-rays caused more damage than either agent alone at all testing times from 1 to 9 months. During the first month the response after the combined treatment was the same as for drug alone (Fig. 16.10).

A comparison of dose–response curves for renal damage at 9 months after X-rays alone or X-rays plus cisplatin demonstrated DEFs of 1.1–1.3, depending on the timing and sequence of the two agents (Fig. 16.11). Independent toxicities probably accounted for most of these effects but, since renal damage was most severe when the drug was given at short intervals before or after irradiation, some modification of the X-ray response could not be ruled out from these single dose studies.

To determine whether cisplatin influenced SLD repair in kidneys, a series of fractionated experiments were carried out. Irradiations (2, 4, 10, 20 or 30 fractions) were given in the first and fourth weeks, with a 2-week rest period in the middle of treatment. Cisplatin (4 mg/kg) was injected 30 minutes before the first fraction of each week, giving a total drug dose of 8 mg/kg. Fig. 16.12 illustrates dose–response curves for renal damage at 31 weeks after treatment. For all fractionation schedules tested there was more damage after the combined

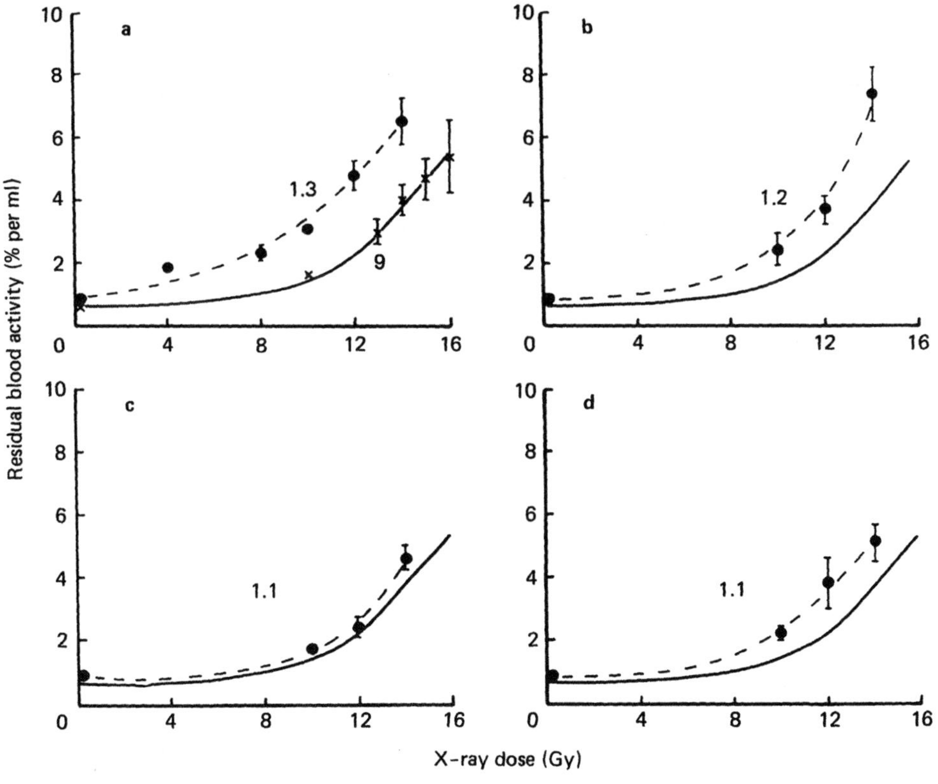

Fig. 16.11. Dose–response curves for kidney damage 9 months after X-rays alone (*continuous line*) or cisplatin (6 mg/kg) (*dashed line*) given before X-rays. DEFs are written beside each pair of curves. **a** Cisplatin ½ hour before irradiation; **b** cisplatin 1 day before irradiation; **c** cisplatin 3 weeks before irradiation; **d** cisplatin 4 weeks before irradiation. (After Stewart et al. 1986.)

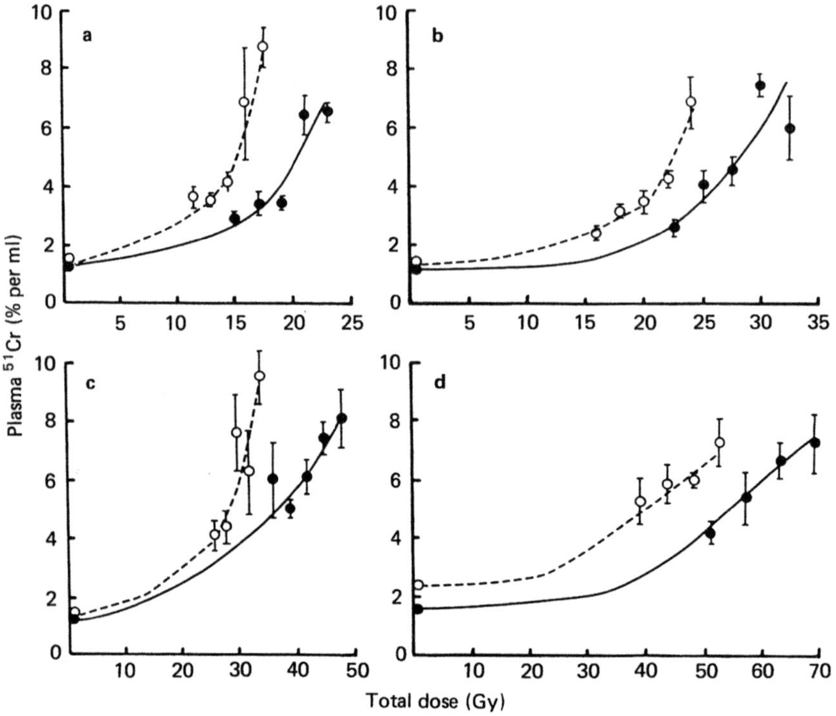

Fig. 16.12. Dose–response curves for renal damage 31 weeks after the start of fractionated treatments with X-rays alone (*filled symbols*) or X-rays plus cisplatin (2 × 4 mg/kg: *open symbols*). Each data point represents the mean of 4–6 mice ±1 SEM. **a** 2 fractions; **b** 4 fractions; **c** 10 fractions; **d** 30 fractions.

treatment than after X-rays alone, with DEFs of 1.3. There was no trend for an increase in DEF with increasing fractionation, as would be expected if the drug caused an inhibition of repair of X-ray damage.

These fractionation data were analysed in terms of an LQ model in order to determine α/β ratios for renal damage after X-rays plus cisplatin. Fig. 16.13 illustrates reciprocal isoeffective doses for the different schedules plotted as a function of the X-ray dose per fraction at a testing time of 30 weeks after the start of the treatment. The straight line fits through the data indicate the applicability of the LQ model for these data, giving α/β ratios (back-extrapolation of the curves to the x-axis) of 3.2 ± 0.2 Gy and 2.3 ± 0.3 Gy for X-rays alone and the combined treatments respectively at this testing time. Repeated tests demonstrated that the α/β ratios for both X-rays alone and the combined treatment varied between 2 and 3 Gy, with no consistent trend for higher, or lower, α/β ratios after the combined treatment. The data points for the combined treatments fall on a separate line from the data for X-rays alone, due to the component of damage from drug alone. If the X-ray dose–response relationship had been modified by cisplatin then a significant change in the α/β ratio would be predicted, but this was not observed. These analyses strongly

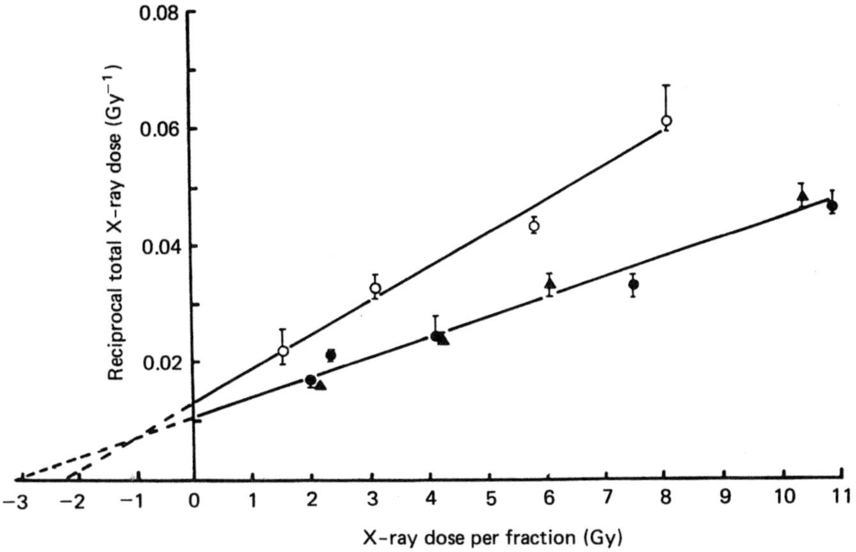

Fig. 16.13. Reciprocal total dose versus dose per fraction at 29–31 weeks from the start of treatment with X-rays alone (*filled symbols*) or X-rays plus cisplatin (*open symbols*). Isoeffective doses were calculated from dose–response curves at a level of 6% residual plasma ^{51}Cr-EDTA. Errors are ±1 SEM, obtained from envelopes of errors on the dose–response curves. ▲, X-ray-only data from a previous experiment giving 2–30 fractions equally spaced in a total treatment time of 1 month; ●, concurrent X-ray-only data with irradiations (2–30 fractions) in the first and fourth week of a 4-week treatment period; ○, X-rays plus cisplatin (2 × 4 mg/kg, given ½ hour before the first X-ray dose in weeks 1 and 4).

imply that radiosensitivity of the kidneys is not modified by cisplatin and that there is no inhibition of SLD repair by the drug. The increased renal damage observed after combined treatments with X-rays and cisplatin is therefore probably the result of independent toxicities.

The increase in renal damage which was observed when cisplatin and X-rays were given in close sequence (within 1 week) was quite modest, with maximum DEF values of 1.3. In a series of retreatment studies, however, mice with previously irradiated kidneys were found to be extremely sensitive to subsequent treatment with cisplatin (Stewart et al. 1987). The LD_{50} for drug given 6 months after renal irradiation (10–12 Gy) was reduced to about 7 mg/kg compared with 14 mg/kg in unirradiated mice. Renal function at the time of drug administration was near normal in all these animals but doses of 4–8 mg/kg cisplatin given to previously irradiated mice caused a rapid and unexpectedly severe deterioration in renal function (Fig. 16.14). Low doses of renal irradiation can therefore markedly decrease tolerance to subsequent cisplatin treatment. It is unlikely that these results can be fully explained in terms of independent toxicities, but the mechanisms involved are, as yet, poorly understood. It is possible that altered pharmacokinetics of the cisplatin in previously irradiated kidneys could explain some of the increased sensitivity. It is noteworthy that the reverse sequence, i.e. renal irradiation given 6 months after previous chemotherapy, did not result in any increased damage compared with irradiation alone.

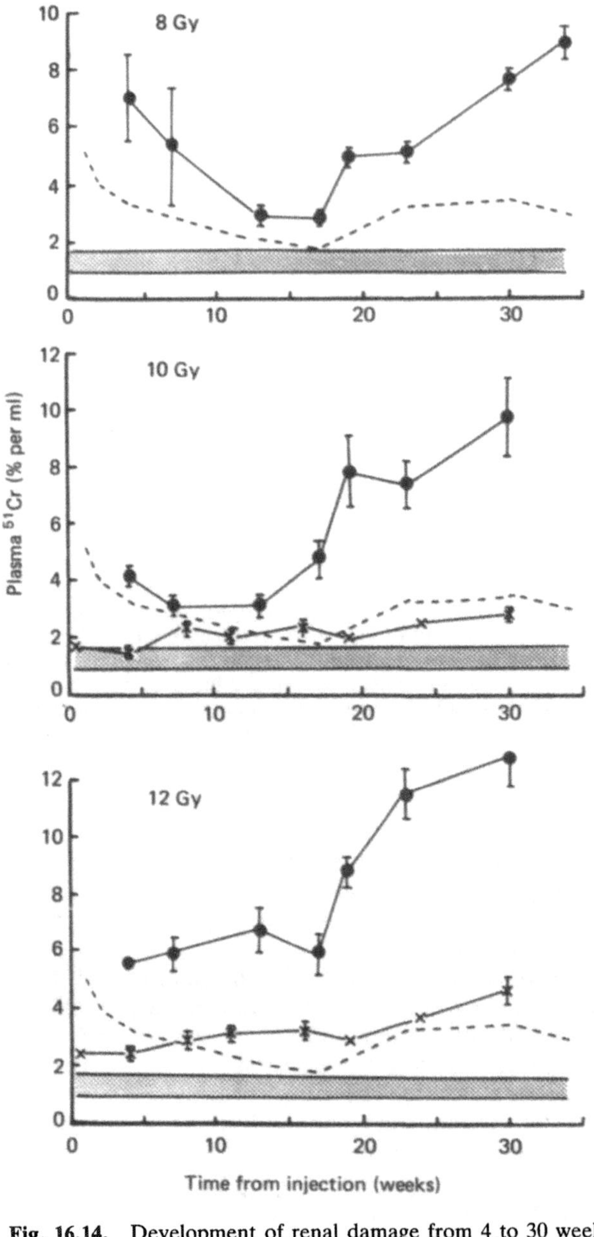

Fig. 16.14. Development of renal damage from 4 to 30 weeks after a dose of 6 mg/kg cisplatin, given 6 months after renal irradiation (●). The responses to drug alone (*dashed line*) and 10–12 Gy X-rays alone (×) are also shown for an equivalent testing period, i.e. from 6 months after irradiation for a period of 30 weeks. The *hatched area* shows the response of control (untreated) mice ±1 SEM. (After Stewart et al. 1987.)

Conclusions

In summary, these studies on normal tissue showed that cisplatin and X-ray treatments given in close sequence caused modest increases in intestinal and renal damage compared with X-rays alone. These effects could largely be explained by independent toxicities, with the possibility of some inhibition of SLD repair in the intestine. There was no increase in late rectal stenosis, acute skin reactions or lip mucosal reactions after combined cisplatin and irradiation. None of these studies demonstrated a large sensitization of normal tissue radiation damage by cisplatin and a comparison with the tumour results (see above) would therefore indicate an overall therapeutic gain. Animals which had previously received renal irradiation were, however, found to have very poor tolerance to cisplatin given several months after irradiation, suggesting that such sequential treatments should be approached with extreme caution.

Clinical Studies

The aim of combined radiotherapy and chemotherapy in patients is primarily to improve survival, but also to achieve better local control and thus avoid both mutilating surgical procedures and patients dying with uncontrollable locoregional recurrences (Bartelink 1987). Promising pilot studies in patients with malignant head and neck tumours, and even widespread advertising, have suggested that this combined treatment regimen is successful, with claims that high response rates, complete remissions and significant tumour reduction can be obtained (Tannock and Brownman 1986). These suggestions were based on non-randomized trials, however, and hardly any long-term results are available from randomized clinical trials to confirm improvement in survival (Tannock and Brownman 1986). Data on the possible detrimental effect of combined treatment are even more scarce.

In this section the available data from clinical trials will be described with regard to a possible gain in local control and survival and the side-effects such as late radiation injuries and induction of second malignancies.

Cytotoxic Drugs as Radiosensitizers

In several trials cytotoxic drugs have been given during the irradiation period to increase the tumour cell kill, with the expectation that they will also act as radiosensitizers. One of the first drugs tested, 5-fluorouracil, appeared to improve the local tumour control in patients with oral cavity tumours (Lo et al. 1976). Other drugs such as hydroxyurea and nitrosoureas have also been tested, but only a small increase in control, for example in brain tumours, has been demonstrated (Reagan et al. 1976). Bleomycin has been used in a number of randomized trials and in two such trials an improvement in local control was obtained (Shanta and Krishnamurhti 1980; Fu et al. 1987). Despite these two positive findings, no beneficial effect could be shown in two larger European studies where bleomycin was used as a radiosensitizer (Cachin et al. 1977;

Vermund et al. 1985; Eschwege et al. 1986). The potential effect of mitomycin C as a hypoxic cell sensitizer has also been investigated in patients with head and neck tumours. Improvement of local control was observed, with a modest increase in survival (Weissberg et al. 1986), but, as in the bleomycin study (Fu et al. 1987), the number of evaluable patients per treatment arm was small. In addition, patients with different primary tumour locations had been entered in the trial and radiotherapy was followed by surgery in some patients. The value of mitomycin C as a radiosensitizer, therefore, still remains unclear. The combination of 5-fluorouracil and mitomycin C has been used in patients with anal carcinoma. Improvement of local control and the avoidance of a colostomy was achieved in the majority of patients (Cummings et al. 1984; Papillon 1986). The possible beneficial effect of combined therapy in this tumour site is now under further study in randomized trials of the EORTC and RTOG groups.

Cisplatin, shown in animal experiments to be a radiosensitizer as described earlier in this chapter, has recently been explored in phase I and II studies (Dewit et al. 1985c; Schaake-Koning et al. 1986). These studies are now being followed by randomized phase III trials in patients with brain, head and neck, oesophagus, lung and bladder cancer. No results are available yet due to insufficient follow-up time.

Radiotherapy and Multi-drug Chemotherapy Regimens

The sequential administration of radiotherapy and chemotherapy is given to obtain independent cell killing. In most trials chemotherapy was given after treatment of the primary tumour with radiotherapy and/or surgery. Despite the biologically sound rationale of adjuvant chemotherapy for the eradication of microscopic metastases, such treatments have not fulfilled their early promise. The entire concept and its biological underpinnings are now under re-evaluation (Carter 1986). Two alternative approaches have been proposed to improve the efficacy of combined treatment. The first is to alternate radiotherapy and chemotherapy to allow the normal tissues to recover from the specific side-effects of each treatment modality (Looney et al. 1985). This approach resulted in a small improvement in the survival of a subgroup of patients with Hodgkin's disease (Hoppe et al. 1979; Tubiana et al. 1985); and in small cell lung cancer a reduced local recurrence rate in the chest was observed with no change in survival (Salazar and Creech 1980; Perez et al. 1984; Maurer et al. 1985). The recent CALGB study (Perry et al. 1987) also demonstrated an improved survival rate when radiotherapy was added to chemotherapy, although the number of patients living longer than 3 years was small. It is therefore likely that in this study a short prolongation of survival cannot be distinguished from a long-lasting improvement in the survival rate.

The second alternative is to give chemotherapy first to reduce the local tumour burden or to avoid a delay of systemic treatment in patients with a high chance of microscopic distant metastases. This approach was explored by the group at Wayne State University for patients with head and neck tumours. They administered a combination of 5-fluorouracil and cisplatin as induction chemotherapy before a course of radiotherapy (Ensley et al. 1984). A high response rate was found, with a correlation between patients who responded well to chemotherapy and long-term local control after radiotherapy. Preliminary results of two randomized trials did not, however, confirm the good results of the pilot studies, and even suggested that a detrimental effect on survival may result due to the delay in the treatment of the primary tumour. (Martin et al. 1986; Haas et al. 1986).

Side-effects of the Combination of Radiotherapy and Chemotherapy

Apart from their own specific toxicity, some cytotoxic drugs enhance radiation damage in normal tissues in patients. This can occur during the radiation period, for example with acute skin and mucosal reactions. The latter phenomenon has been observed when bleomycin was given during the irradiation period (Cachin et al. 1977; Eschwege et al. 1986). The increased oesophagitis observed in patients correlates well with results from animal experiments in which mouse lip mucosal reactions induced by irradiation were significantly increased when the animals received concomitant bleomycin infusions (Vanuytsel et al. 1986). In other clinical studies cisplatin given during the irradiation period also caused more severe oesophagitis reactions in patients with inoperable lung cancer (C. Schaake, personal communication).

Information on enhancement of late radiation damage by cytostatic agents is scarce. Administration of 5-fluorouracil was associated with more late eye complications when it was added to irradiation in patients with sinus maxillary tumours (Chan and Shukovsky 1976). Bleomycin, on the other hand, did not increase late radiation damage in patients who had been treated for oropharynx carcinoma (Eschwege et al. 1986). Most randomized trials, however, have not used criteria for assessing late normal tissue damage.

Another point for concern is the possible increase in secondary malignancies after combined therapy. Several studies have convincingly shown that the use of alkylating agents in patients with non-Hodgkin's lymphoma increases the risk of leukaemia (Gomez et al. 1982; Greene et al. 1983; Pedersen-Bjergaard et al. 1985). This finding is consistent with studies on the risk of secondary leukaemia following treatment of Hodgkin's disease and ovarian cancer (Pedersen-Bjergaard et al. 1985). In most studies the induction of second malignancies is largely attributed to the chemotherapy. Radiotherapy did not seem to have an additive effect on the occurrence of these leukaemias. Recently there have also been some suggestions that with longer follow-up more solid tumours are being found in the combined treatment groups (Somers et al. 1987).

Future Directions

There is still a lack of understanding as to why certain chemotherapy drugs are radiosensitizers and how they act. This stems partly from the as yet rather hazy understanding of the factors determining intrinsic radiosensitivity, plus an equally hazy appreciation of the molecular lesion(s) responsible for cell death from a particular drug. Increased knowledge in this area is obviously of fundamental importance in first understanding and then improving combination treatments with the two agents. An indication of how this is being tackled for cisplatin was discussed above.

Looking to the future, it is hoped that a correlation can be shown, for example, between radiosensitization by cisplatin and the degree of DNA adduct formation of the type that can readily be detected and quantified in tissue sections by the use of specific antiserum (Terheggen et al. 1987). This would provide a relatively straightforward method of testing individual tumours in patients for sensitivity to cisplatin, i.e. adduct detection in a biopsy taken shortly after the first course of the drug. Such antibody detection methods could well be developed to a variety of DNA-binding drugs and to drugs which alter surface properties.

Another approach for improving combined modality therapy is to find ways of selectively increasing drug concentrations in tumour without concomitant increases in sensitive normal tissues, that is, drug targeting. One relatively novel approach that has been receiving attention recently is that of intratumoral injection of the drug held in a slow-release system (Yu et al. 1987; Howes et al. 1987; Begg et al. 1987). Potential advantages of such an approach are the relatively high tumour to normal tissue drug concentration ratios achievable plus the persistence of drug in the tumour over an extended fractionated radiotherapy course (the latter mediated through the choice of a suitable slow-release vehicle). Improvements in techniques for monitoring the placement of intratumoral catheters and probes extend the possibilities for this type of therapy beyond its use for superficial tumours. This may ultimately prove to be a better drug/radiosensitizer targeting approach for radiotherapy than the use of liposomes, monoclonal antibodies, etc. Much fundamental animal work with such systems, however, has still to be done.

References

Bartelink H (1987) Combined modality treatment for the primary tumour. National Cancer Institute Monographs (in press)

Bartelink H, Kallman RF, Rapachietta D, Hart GAM (1986) Therapeutic enhancement in mice by clinically relevant dose and fractionation schedules of cis-diamminedichloroplatinum (II) and irradiation. Radiother Oncol 6:61–74

Begg AC, van der Kolk PJ, Dewit L, Bartelink H (1986) Radiosensitization by cisplatin of RIF-1 tumour cells in vitro. Int J Radiat Biol 50:871–884

Begg AC, Bartelink H, Stewart FA, Brown DM, Luck EE (1987) Improvement of differential toxicity between tumor and normal tissues using intratumoral injections with and without a slow drug release matrix system. National Cancer Institute Monographs (in press)

Bergerat JP, Barlogie B, Göhde W, Johnston DA, Drewinko B (1979) In vitro cytokinetic response of human colon cancer cells to cis-diamminedichloroplatinum (II). Cancer Res 39:4356–4363

Cachin Y, Jortay A, Sancho H et al. (1977) Preliminary results of a randomized EORTC study comparing radiotherapy and concomitant bleomycin to radiotherapy alone in epidermoid carcinomas of the oropharynx. Eur J Cancer 13:1389–1395

Carde P, Laval F (1981) Effects of cis-diamminedichloroplatinum (II) and X-rays on mammalian cell survival. Int J Radiat Oncol Biol Phys 7:929–933

Carter SK (1986) Adjuvant chemotherapy of cancer. A review of its current status. Drugs 31(4):337–367

Chan RC, Shukovsky LJ (1976) Effects of irradiation on the eye. Radiology 120:673–675

Cummings BJ, Keane TJ, Thomas GM, Harwood AR, Rider WD (1984) Results and toxicity of the treatment of anal canal carcinoma by radiation therapy or radiation therapy and chemotherapy. Cancer 54:2062–2068

Dewit L, Oussoren Y, Bartelink H (1985a) Dose and time effects of cis-diamminedichloroplatinum (II) and radiation in mouse duodenal crypts. Radiother Oncol 4:363–371

Dewit L, Begg AC, Kohler Y, Stewart FA, Bartelink H (1985b) Influence of cis-diamminedichloroplatinum (II) on mouse duodenal crypt cell survival after multifraction X ray treatment. Int J Radiat Oncol Biol Phys 11:1809–1816

Dewit L, Bartelink H, Rumke Ph (1985c) Concurrent cis-diamminedichloroplatinum (II) and radiation treatment for melanoma metastases: a pilot study. Radiother Oncol 3:303–309

Dewit L, Oussoren Y, Bartelink H (1987) Early and late damage in the mouse rectum after irradiation and cis-diamminedichloroplatinum (II). Radiother Oncol 8:57–69

Douple EB, Richmond RC (1980) A review of interactions between platinum coordination complexes and ionizing radiation: implications for cancer therapy. In: Prestayko AW, Crooke ST, Carter SK (eds) Cisplatin – current status and new developments. Academic Press, New York, pp 125–147

Douple EB, Richmond RC (1982) Enhancement of the potential of radiotherapy by platinum drugs in a mouse tumor. Int J Radiat Oncol Biol Phys 8:501–503

Dritschilo A, Piro A, Kelman AD (1979) The effect of cis-platinum on the repair of radiation damage in plateau phase Chinese hamster (V-79) cells. Int J Radiat Oncol Biol Phys 5:1345–1349

Ensley JF, Jacobs JR, Weaver A et al. (1984) Correlation between response to cisplatinum-combination chemotherapy and subsequent radiotherapy in previously untreated patients with advanced squamous cell cancers of the head and neck. Cancer 54:811–814

Eschwege F, Sancho-Garnier H, Cachin Y, Jortay A, Madelain M, Desaultry A (1986) Bleomycin and radiation therapy in squamous cell carcinoma of the oropharynx. Long term results of a phase III clinical trial. EORTC Meeting, Williamsburg, Virginia, 28 September–1 October 1986 (abstr):2/9

Fu KK, Lam KN, Rayner PA (1985) The influence of time sequence of cisplatin administration and continuous low dose rate irradiation (CLDRI) on their combined effects on a murine squamous cell carcinoma. Int J Radiat Oncol Biol Phys 11:2119–2124

Fu KK, Phillips TL, Silverberg IJ et al. (1987) Combined radiotherapy and chemotherapy with bleomycin and methotrexate for advanced inoperable head and neck cancer: update of a Northern California Oncology Group randomized trial. J Clin Oncol 5:1410–1418

Gomez GA, Aggarwal KK, Han T (1982) Post therapeutic acute malignant myeloproliferative syndrome and acute non-lymphocytic leukemia in non-Hodgkin's lymphoma: correlation with intensity of treatment. Cancer 50:2285–2288

Greene MH, Young RC, Merrill JM, DeVita VT (1983) Evidence of a treatment dose response in acute nonlymphocytic leukemias which occur after therapy of non-Hodgkin's lymphoma. Cancer Res 43:1891–1898

Haas C, Anderson T, Byhardt R et al. (1986) Randomized neo-adjuvant study of S-fluorouracil and cis-platinum for patients with advanced resectable head and neck squamous cancer. Proc Am Assoc Cancer Res 27:185

Harder HC, Rosenberg B (1970) Inhibitory effects of anti-tumor platinum compounds on DNA, RNA and protein synthesis in mammalian cells in vitro. Int J Cancer 6:207–216

Hoppe RT, Portlock CS, Glatstein E, Rosenberg SA, Kaplan HS (1979) Alternating chemotherapy and irradiation in the treatment of advanced Hodgkin's disease. Cancer 43:472–481

Howes AE, Herman TS, Montoya VP, Luck EE, Brown DM (1987) Effect of matrix associated chemotherapy in combination with X-rays in vivo. National Cancer Institute Monographs (in press)

Landuyt W, Ang KK, van der Schueren E (1986) Combinations of single dose and fractionated treatments of cis-diamminedichloroplatinum (II) and irradiation: effect on mouse lip mucosa. Br J Cancer 54:579–586

Lelieveld P, Scoles MA, Brown JM, Kallman RF (1985) The effect of treatment in fractionated schedules with the combination of x-irradiation and six cytotoxic drugs on the RIF-1 tumor and normal mouse skin. Int J Radiat Oncol Biol Phys 11:111–121

Lo TC, Wiley AL, Ansfield FJ et al. (1976) Combined radiation therapy and 5-fluorouracil for advanced squamous cell carcinoma of the oral cavity and oropharynx: a randomized study. Am J Roentgenol 126:229–235

Looney WB, Hopkins HA, Carter WH (1985) Solid tumor models for the assessment of different treatment modalities. XXIII. A new approach to the more effective utilization of radiotherapy alternated with chemotherapy. Int J Radiat Oncol Biol Phys 11:2105–2117

Luk KH, Ross GY, Phillips TL, Goldstein LS (1979) The interaction of radiation and cis-diamminedichloroplatinum (II) in intestinal crypt cells. Int J Radiat Oncol Biol Phys 5:1417–1420

Martin M, Mazeron JJ, Glaubiger D et al. (1986) Neo-adjuvant polychemotherapy of head and neck cancer: preliminary results of a randomized study. Proc Asco 5:141

Maurer LH, Pajak T, Eaton W et al. (1985) Combined modality therapy with radiotherapy, chemotherapy, and immunotherapy in limited small-cell carcinoma of the lung: a phase III Cancer and Leukaemia Group B study. J Clin Oncol 3:969–976

Murthy AK, Rossot AH, Anderson KM, Hendrickson FR (1979) Cytotoxicity and influence on radiation dose response curve of cis-diamminedichloroplatinum II (cis-DDP). Int J Radiat Oncol Biol Phys 5:1411–1415

Overgaard J, Khan AR (1981) Selective enhancement of radiation response in a C3H mammary carcinoma by cisplatin. Cancer Treat Rep 65:501–503

Papillon J (1986) Current therapeutic concepts of management of carcinoma of the anal canal. Fifth Annual Meeting of the European Society for Therapeutic Radiology and Oncology, Baden-Baden

Pedersen-Bjergaard J, Ersbøll J, Sørensen HM et al. (1985) Risk of acute non-lymphocytic leukemia and preleukemia in patients treated with cyclophosphamide for non-Hodgkin's lymphomas. Ann Intern Med 103:195–200

Perez CA, Einhorn L, Oldham RK et al. (1984) Randomized trial of radiotherapy to the thorax in limited small-cell carcinoma of the lung treated with multi-agent chemotherapy and elective brain irradiation: a preliminary report. J Clin Oncol 2:1200–1208

Perry MC, Eaton WL, Propert KJ et al. (1987) Chemotherapy with or without radiation therapy in limited small-cell carcinoma of the lung. N Engl J Med 316:912–918

Phillips TL (1979) Rationale for the selection of combined treatment schedules using fractionated radiation and chemotherapy. In: Moore M (ed) Advances in medical oncology, research and education. Pergamon Press, New York, pp 183–190

Phillips TL, Wharam MD, Margolis LW (1975) Modification of radiation injury to normal tissues by chemotherapeutic agents. Cancer 35:1678–1684

Poirier MC, Lippard SJ, Zwelling LA et al. (1982) Antibodies elicited against cis-diamminedichloroplatinum (II)-modified DNA are specific for cis-diamminedichloroplatinum (II)–DNA adducts formed in vivo and in vitro. Proc Natl Acad Sci USA 79:6443–6447

Reagan TJ, Bisel HE, Childs DS, Layton DD, Rhoton AL, Taylor WF (1976) Controlled study of CCNU and radiation therapy in malignant astrocytoma. J Neurosurg 44:186–190

Salazar OM, Creech RH (1980) "The state of the art": toward defining the role of radiation therapy in the management of small cell bronchogenic carcinoma. Int J Radiat Oncol Biol Phys 6:1103–1117

Schaake-Koning CCE, Bartelink H, Schuster-Uitterhoeve L et al. (1986) Radiotherapy combined with cis-diamminedichloroplatinum (c-DDP) as a radio enhancer for inoperable non small cell lung cancer. Int J Radiat Biol 50:366–367

Shanta V, Krishnamurhti S (1980) Combined bleomycin and radiotherapy in oral cancer. Clin Radiol 31:617–620

Shenken LL, Burholt DR, Hagemann RF, Kovacs CJ (1979) Combined modality oncotherapies. Front Radiat Ther Oncol 13:82–101

Somers R, VanLeeuwen FE, Taal BG, Koster B, Huisman SJ (1987) Secondary malignancies after Hodgkin's disease in the Netherlands Cancer Institute. Third international conference on malignant lymphomas, Lugano, 10–13 June (abstr)

Steel GG, Peckham MJ (1979) Exploitable mechanisms in combined radiotherapy–chemotherapy: the concept of additivity. Int J Radiat Oncol Biol Phys 5:85–91

Stewart F, Bohlken S, Begg AC, Bartelink H (1986) Renal damage in mice after treatment with cisplatinum alone or in combination with X-irradiation. Int J Radiat Oncol Biol Phys 12:927–933

Stewart FA, Luts A, Begg AC (1987) Tolerance of previously irradiated mouse kidneys to cis-diamminedichloroplatinum (II). Cancer Res 47:1016–1021

Tannock IF, Brownman G (1986) Lack of evidence for a role of chemotherapy in the routine management of locally advanced head and neck cancer. J Clin Oncol 4:1121–1126

Terheggen PMAB, Scherer E, Floot B, Begg AC, Fichtinger-Schepman MJ, DenEngelse L (1987) Immunocytochemical detection of interaction products of cis-diamminedichloroplatinum II (cis-DDP) and cis-diammine (1,1-cyclobutanedicarboxylato)platinum II (c-DCBA) with DNA in rodent tissue sections. Cancer Res (in press)

Tubiana M, Arriagada R, Cosset JM (1985) Sequencing of drugs and radiation. The integrated alternating regimen. Cancer 55:2131–2139

Vanuytsel L, Feng Y, Landuyt W, Leer JW, van der Schueren E (1986) The combined effect of bleomycin and irradiation on mouse lip mucosa. Influence on the accumulation and repair of sublethal damage during fractionated irradiation. Radiother Oncol 6:267–273

Vermund H, Kaalhuis O, Winther F, Trausio J, Thorud E, Harang R (1985) Bleomycin and radiation therapy in squamous cell carcinoma of the upper aero-digestive tract: a phase III clinical trial. Int J Radiat Oncol Biol Phys 11:1877–1886

Von der Maase H (1984a) Interactions of radiation and adriamycin, bleomycin, mitomycin C or cis-diamminedichloroplatinum (II) in intestinal crypt cells. Br J Cancer 49:779–786

Von der Maase H (1984b) Effect of cancer chemotherapeutic drugs on the radiation-induced skin reactions in mouse feet. Br J Radiol 57:697–707

Weissberg JB, Son YH, Papac RJ et al. (1986) Controlled clinical trial of mitomycin C (MC) as an adjunct to radiotherapy (RT) in head and neck cancer. EORTC Meeting, Williamsburg, Virginia, 28 September–1 October 1986 (abstr):2/12

Withers HR, Elkind MM (1970) Microcolony survival assay for cells of mouse intestinal mucosa exposed to radiation. Int J Radiat Biol 17:261–267

Yu NY, Conley FK, Luck EE, Brown DM (1987) Response of murine tumors to matrix associated cis-DDP intratumoral implants. National Cancer Institute Monographs (in press)

Ziegler W (1986) Cisplatin, a sensitizer of hypoxic mammalian cells? J Cancer Res Clin Oncol 111 [Suppl]:17

17 Biological Aspects of Hyperthermia

S. B. Field

Introduction

For more than a century physicans have been aware that prolonged high fever can lead to regression of tumours. However, the development of a rationale for the use of heat in treating malignant disease began in the 1960s with a systematic study of its biological effects. A surge of interest in the field then followed, with detailed studies of the biology and physiology of hyperthermia, development of improved methods of heat delivery and temperature measurement and a substantial number of clinical studies leading now to the possibility of randomized controlled trials.

Hyperthermia can be used as a cell killing modality in its own right, although, in practice, it is normally added to one of the conventional forms of treatment, such as radiotherapy. Alternatively it can be used to enhance the effect of either radiotherapy or some forms of chemotherapy.

Rationale for the Use of Hyperthermia in Cancer Therapy

The reasons why heat may be a useful anti-cancer treatment depend mainly on the fact that blood flow in tumours, or at least in some regions of tumours, is more disorganized and often sluggish compared with that in normal tissues (Reinhold and Endrich 1986). This will have two important consequences. First, tumours will have a poorer cooling mechanism than normal tissues and will, therefore, become hotter in a localized heating field. This has been observed many times in both animals and man. It is also known that substantial numbers of tumour cells are likely to be sufficiently far from blood vessels to become radiobiologically hypoxic and, hence, radioresistant. These hypoxic cells are at least as sensitive to hyperthermia as are well oxygenated cells; furthermore, hypoxic tumour cells have a strong tendency towards anaerobic glycolysis with the consequent production of lactic acid. Thus, tumours (or regions of tumours) are often at a lower pH than normal tissues (Gullino et al. 1964) and it is known that reducing the pH makes cells highly heat sensitive (Gerweck 1977). In addition, cellular reponse to heat is dramatically affected by the nutrient environment such as serum proteins, high sensitivity being equated with nutrient deficiency (Hahn 1982). In both cases the differential response is substantial.

There are reports which show neoplastic cells to be more heat sensitive than their normal counterparts. However, careful studies of the heat sensitization of normal and transformed cells have not generally indicated a significantly enhanced sensitivity for the transformed cell.

Since the sensitivity of cells as a function of their position in the cell cycle is opposite for X-rays and hyperthermia, it is possible that combining the two modalities might lead to a therapeutic advantage.

Response to Hyperthermia of Cells and Tissues

Temperatures of around 41 °C or greater can be lethal to mammalian cells. However, the targets responsible for cell death have not been unequivocally identified amongst the multitude of cellular changes that have been reported to occur after hyperthermia (Leeper 1985). During DNA synthesis heated cells appear to die as the result of chromosomal injury. However, DNA synthesis takes a very small proportion of the total cell cycle time in many tissues. At other times during the cell cycle the primary target for hyperthermia is thought to be membrane – either the plasma membrane or cytoskeleton. Membrane injury causes cells to die and lyse rapidly during interphase. In contrast, after equally lethal doses of ionizing radiation cells do not die until division is attempted. Thus many tissues, especially those that are slowly dividing, respond far more quickly to hyperthermia than to ionizing radiation. Heat also kills cells that are unaffected by ionizing radiations, such as the post-mitotic cells on intestinal villi, and the effects become noticeable almost immediately after heating (Hume et al. 1983).

The pathological effects of hyperthermia in various normal tissues have been reviewed by Fajardo (1984). In general, the response is similar to but less severe than a thermal burn. Depending on its magnitude, hyperthermia may be followed by oedema, focal haemorrhage, granulocytic infiltration and even necrosis within the first day or so. Later changes include loss of blood vessels, necrosis, monocyte infiltration and fibrosis. Most observations indicate that once the acute response to heat has healed, there are no late effects, but we cannot yet be quite certain of this.

There is no doubt that damage to supporting tissues, particularly the vasculature, plays an extremely important role in heat injury to both tumours and normal tissues. Of particular importance is the observation made on transplanted animal tumours that vascular breakdown and stasis occur early, that is at lower temperatures or at earlier times, than in normal tissues. The topic of effects of heat on the vasculature and microcirculation has been reviewed recently by Reinhold and Endrich (1986).

Lysosomal damage has also been reported by several groups as being associated with heat injury; effects include increased lysosomal enzyme activity, increased membrane permeability and a higher concentration of lysosomes. However, it is not yet clear whether these changes are the consequence of tissue damage or the cause (Hahn 1982).

The effects of heat have been quantified in many different cell lines, normal tissues and experimental tumours. In normal tissues it is observed that at any given temperature once the time of heating is sufficiently long to reach the threshold of injury, only a small increase in heating time is required to increase the probability of necrosis to 100%. This increase is approximately 20% in heating time or a rise in temperature of only 0.5 deg C. One consequence of this effect is that small differences in tissue temperature, which are very likely to occur in clinical treatments, may lead to marked differences in biological responses if the heat treatment is close to the damage threshold. Therefore, great care must be taken to monitor temperatures at many points. On the other hand, if tumours are only slightly more sensitive, or at a slightly higher temperature than the surrounding normal tissues, then hyperthermia could result in a clear therapeutic advantage.

It is important to understand the relationships between time of heating and temperature to produce a given effect. Many data of this kind are now available and it is found that, in general, the slopes of such curves are fairly similar. An analysis of all such data, in vivo and in vitro, has shown the mean time change equivalent to a temperature difference of 1 deg C is equal to a factor of 2.1 ± 0.07 above a transition point and 6.4 ± 0.4 below the transition, which itself normally occurs between 42 and 43 °C. However, it is also found that there are very large differences in tissue sensitivity (Field and Morris 1983).

It has been shown that if the blood supply to a tissue is occluded the tissue becomes far more heat sensitive. There is a shift in the complete isoeffect curve, presumably due to the resulting acidification. If tumours, or at least regions within tumours, are at a reduced pH then the heat response might be represented by the more sensitive curve. Also, the transition in isoeffect appears to be abolished by occlusion of the blood supply. Thus, the maximum differential might be achieved by heating at relatively low temperatures for long times, although this may not be very practical clinically (Field and Morris 1985).

Thermotolerance

An important and very general finding is that heat treatment leads to a resistance to further heating – a phenomenon known as thermotolerance (Henle and Dethlefsen 1978). Two types of thermotolerance have been identified: that which develops during prolonged heating below a critical temperature (about 42 °C) and that which develops between individual hyperthermia fractions. It is an extremely large effect. For example, the slope of a survival curve may be decreased by as much as a factor of 15. Alternatively, in several tissues the time of heating required to produce a given effect is increased by a factor of 2–4 when the tissue is maximally thermotolerant. The effect is transient, the decay depending on the particular cell type or tissue.

Thermotolerance in vivo and in vitro follow similar patterns, strongly suggesting that it is a cellular, rather than a physiological, phenomenon. Reduced pH has been shown to cause a reduction in thermotolerance, so that thermotolerance may be less in some tumours (or regions of tumours) than in normal tissues. A few studies, however, indicate that thermotolerance in animal tumours lies within the range of results for normal tissues. Thus, the experimental studies to date have given no clear indication that tumours do contain cells which have a reduced capacity to develop thermotolerance and there is no basis for optimism that differences between the development of thermotolerance in normal tissues and tumours may lead to a therapeutic gain.

The time course of thermotolerance is complex. The consensus at present is that it is preferable, clinically, to avoid thermotolerance effects, if possible, by waiting sufficiently long between hyperthermic treatments for the effects to have decayed (Field 1985).

The molecular basis for thermotolerance is not understood, but there are numerous reports that hyperthermia results in the synthesis of a specific set of proteins known as heat stress or heat shock proteins. There is some evidence that thermotolerance is related to such protein synthesis (Burdon 1985).

An operational model which helps in formulating a picture of thermotolerance was proposed by Li and Hahn (1980). It involves three separate phases. The first is the induction or trigger, which can occur at any temperature. The second is the development of thermotolerance, which can only occur at temperatures below about 41–42 °C, depending on the type of cell, so that with two heat treatments it is necessary to return the cells (or tissues) to temperatures below 41 or 42 °C for thermotolerance to develop. The third phase is decay.

The topic of thermotolerance has often been reviewed, for example by Henle and Dethlefsen (1978), Hahn (1982), Field and Anderson (1982) and Field (1985).

Step-Down Sensitization

If cells or tissues are heated at approximately 43 °C or above, followed by heating at 42 °C or below, then the lower temperature becomes more effective than if given alone (Henle et al. 1978). The effect normally occurs only if the temperature of the first treatment is higher than the transition temperature and that of the second treatment is lower. It has been suggested that step-down sensitization results from abolishing thermal tolerance, although it is perhaps more likely that it is an independent phenomenon. It has recently been suggested that at the higher temperatures more lesions are created than can be fixed, whereas at the lower temperatures it is possible to fix many of the lesions already present, thus increasing the effectiveness of the lower temperatures (Jung 1986). The step-down effect is present immediately after the first heating and decays rapidly.

Interactions Between Hyperthermia and Other Modalities

Chemotherapy

Combining heat with chemotherapeutic drugs may enhance the therapeutic effect of the drug either by simply increasing its uptake or by enhancing sensitivity to it. There exists a range of types of interaction. For example, some compounds, such as adriamycin and bleomycin, are unaffected below a threshold temperature above which there may be a dramatic increase in the drug's toxicity. Other compounds, such as amphotericin B and alcohols, are not toxic at all at physiological temperatures but only become toxic at increased temperatures. A continuously increasing effect with increasing temperature is seen for *cis*-platinum, alkylating agents and nitrosoureas. For yet other compounds, such as methotrexate, there appears to be no interaction with temperature. This topic is treated in some detail by Hahn (1982).

Radiation

Heat causes both an increase in the intrinsic sensitivity of cells to irradiation and a reduction in the capacity of cells to repair both sublethal and potentially lethal injury.

In general, direct thermal injury is observed at times before radiation damage is expressed. Mild heat treatments, which have little effect if given alone, may enhance the response of tissues to ionizing radiation. The response to the combined treatment is qualitatively similar to that following radiation alone. Therefore the effect of a mild heat treatment on the radiation response can be expressed as a thermal enhancement ratio (TER), defined as the dose of radiation required to cause a given response in the absence of hyperthermia divided by that required to cause the same response in combination with hyperthermia. The crucial question is whether or not the TER for tumours is significantly greater than for normal tissues. This has not been found to be the case in the experimental studies. The explanation is as follows:

1. Under the circumstances of heating by means of water baths, as was nearly always the case in experimental studies, the tumours were not hotter than the water; indeed, they often contained regions which were cooler. With clinical treatments this is probably different.

2. Studies in vitro indicate that when heat is used to enhance radiation response, pH is far less important than for direct heat cytotoxicity (Lunec and Parker 1980).

3. There do not appear to be any data on whether nutrient deficiency affects the potential of heat to enhance radiation injury.

4. It appears that the oxygen enhancement ratio for ionizing radiation (about 3 for photons) does not change significantly for the combination of heat and radiation. Thus the relative radioresistance of hypoxic cells is not improved by hyperthermia (Field and Bleehen 1979).

5. Whether neoplasia, per se, is important to direct heat injury is dubious: there do not appear to be data for heat enhancement of radiation damage.

Thus the clinical rationale for using heat to enhance radiation damage is far weaker than that for using heat alone. Other ways must be sought to combine heat with ionizing radiation for it to be clinically effective.

Effects of Sequence and Interval

If heat and radiation are given simultaneously, the effect is maximal. As the separation between the two modalities is increased, the enhancing effect decreases. This is an especially consistent finding if heat is given after X-rays. The interaction in normal tissues is lost when the interval is increased to approximately 4 hours or more. Thus, giving heat approximately 4 hours after radiation results in the two modalities acting independently. The TER for normal tissue is reduced to unity but that for tumour remains greater due to direct toxicity, resulting in a therapeutic advantage (Overgaard 1982).

Re-treatment of Previously Irradiated Tissue

Hyperthermia is often used clinically to treat tumours which have recurred some time after more conventional therapy. Most clinicians have reported that such treatment is normally without complications, although a recall effect of earlier treatment has been reported after radiotherapy to the central nervous system (Douglas et al. 1981).

Several studies on rodent skin and intestine have shown a marked but transient increase in sensitivity to hyperthermia when heat is given at various times after radiation treatment. The time course of this increase in heat sensitivity varies from tissue to tissue, but it appears that, providing the initial radiotherapy treatment is not too severe, thermal sensitivity returns approximately to normal (Law 1987).

Previously irradiated sites may also be re-treated by a combination of hyperthermia and radiotherapy. In such cases there appears to be no change in the ability of a mild heat treatment to enhance radiation injury. Thus the TER is not increased by prior exposure to X-rays (Law et al. 1985).

Thermal Dose

Quantitation of thermal dose in a biologically relevant way is extremely difficult. Even the term dose is not very rigidly defined. Radiation workers have become accustomed to using energy deposited to describe dose of radiation, since this relates clearly to the resulting effect. Pharmacologists use drug concentration or weight, which also relates to effect. With hyperthermia the situation is rather different, the biological response being primarily dependent on the time at an elevated temperature, not on the deposition of energy (Hahn 1982). Thus, if the temperature were fixed and constant, time at the raised temperature would be a perfectly reasonable method of defining a thermal dose. However, the temperature is not fixed and in clinical practice is certainly far from constant.

Gerner (1985) proposed a thermodynamic approach to this problem, based on the Arrhenius equation, i.e. the reaction rate constant $k = A\ e^{-H/RT}$, where H is the activation energy, T is the temperature and R is the gas constant.

Dewey et al. (1977) proposed using a simple empirical formula, based on in vitro observations:

$$\frac{t_2}{t_1} = R^{T_1 - T_2} \qquad \begin{array}{l} R = 2 \text{ for } T > 42.5\ ^\circ\mathrm{C} \\ R \simeq 6 \text{ for } T < 42.5\ ^\circ\mathrm{C} \end{array}$$

This equation provides a means of relating treatments to a given tissue with different temperatures (T) and times of heating (t), but it takes no account of differences in absolute tissue sensitivity. These two approaches are very similar, the time–temperature relationship itself being required to calculate the activation energy. Sapareto and Dewey (1984) proposed that 43 °C should be used as a reference temperature and that all treatments be described as equivalent minutes of heating at 43 °C obtained by integration of the above formula throughout a treatment.

The integration procedure has been modelled experimentally and it was concluded from such studies that the Dewey formula provides a practical and reasonable method of comparing hyperthermal treatments under conditions likely to be met in practice, that is moderate variation about a fairly steady temperature (Field 1987). However, the formula would certainly be inaccurate if there were large variations in temperature resulting in a significant effect of step-down sensitization. Also it must be emphasized that the formula does not address the problem of varying sensitivity throughout a course of fractionated heat treatments. The use of this isoeffect relationship must be seen as an interim solution to the problem of estimating a thermal dose.

Conclusions

It can be seen that there are sound biological reasons for an interest in hyperthermia. Successful clinical implementation will depend on good physics and engineering techniques and appropriate therapeutic design. These are discussed in the next chapter.

References

Burdon RH (1985) Heat shock proteins. In: Overgaard J (ed) Hyperthermic oncology. Taylor and Francis, London, pp 223–230

Dewey WC, Hopwood LE, Sapareto SA, Gerweck LE (1977) Cellular responses to combinations of hyperthermia and radiation. Radiology 123:463–474

Douglas MA, Parks LC, Bebin J (1981) Sudden myelopathy secondary to therapeutic total-body hyperthermia after spinal cord irradiation. N Engl J Med 10:583–585

Fajardo LF (1984) Pathological effects of hyperthermia in normal tissues. Cancer Res 44 [Suppl]:4826s–4835s
Field SB (1985) Clinical implications of thermotolerance. In: Overgaard J (ed) Hyperthermic oncology. Taylor and Francis, London, pp 235–244
Field SB (1987) Studies relevant to a means of quantifying effects of hyperthermia. Int J Hyperthermia 3:291–296
Field SB, Anderson RL (1982) Thermotolerance: a review of observations and possible mechanisms. Nat Cancer Inst Monog 61:193–201
Field SB, Bleehen NM (1979) Hyperthermia in the treatment of cancer. Cancer Treat Rev 6:63–94
Field SB, Morris CC (1983) The relationship between heating time and temperature: its relevance to clinical hyperthermia. Radiother Oncol 1:179–186
Field SB, Morris CC (1985) Experimental studies of thermotolerance *in vivo*. 1. The baby rat tail model. Int J Hyperthermia 1:235–246
Gerner E (1985) Definition of thermal dose; biological isoeffect relationships and dose for temperature-induced cytotoxicity. In: Overgaard J (ed) Hyperthermic oncology. Taylor and Francis, London, pp 245–252
Gerweck LE (1977) Modification of cell lethality at elevated temperatures: the pH effect. Radiat Res 70:224–235
Gullino PM, Clarke SH, Grantham FH (1964) The interstitial fluid of solid tumours. Cancer Res 24:780–794
Hahn GM (1982) Hyperthermia and cancer. Plenum Press, New York
Henle KJ, Dethlefsen LA (1978) Heat fractionation and thermotolerance: a review. Cancer Res 38:1843–1851
Henle KJ, Karamuz JE, Leeper DB (1978) Induction of thermotolerance in Chinese hamster ovary cells by high (45°) or low (40°) hyperthermia. Cancer Res 38:570–574
Hume SP, Marigold JCL, Michalowski A (1983) The effect of local hyperthermia on non-proliferative, compared with proliferative, epithelial cells of the mouse intestinal mucosa. Radiat Res 94:252–262
Jung J (1986) A generalised concept for cell killing by heat. Radiat Res 106:56–72
Law MP (1987) The response of normal tissues to hyperthermia. In: Urano M (ed) Hyperthermia and oncology. VNU Scientific, Netherlands (in press)
Law MP, Ahier RG, Somaia S (1985) Thermal enhancement of radiation damage in previously irradiated skin. Br J Radiol 58:161–167
Leeper DB (1985) Molecular and cellular mechanisms of hyperthermia alone or combined with other modalities. In: Overgaard J (ed) Hyperthermic oncology. Taylor and Francis, London, pp 9–40
Li GC, Hahn GM (1980) A proposed operational model of thermotolerance based on effects of nutrients and the initial treatment temperature. Cancer Res 40:4501–4508
Lunec J, Parker R (1980) The influence of pH on the enhancement of radiation damage by hyperthermia. Int J Radiat Biol 38:567–574
Overgaard J (1982) Influence of sequence and interval on the biological response to combined hyperthermia and radiation. Nat Cancer Inst Monogr 61:325–332
Reinhold HS, Endrich B (1986) Tumour microcirculation as a target for hyperthermia. Int J Hyperthermia 2:111–138
Sapareto SA, Dewey WC (1984) Thermal dose determination in cancer therapy. Int J Radiat Oncol Biol Phys 10:787–806

18 Clinical Hyperthermia: Methods and Results

N. M. Bleehen and G. C. W. Howard

Introduction

Clinical hyperthermia traces its history back to antiquity. The earliest examples, as represented by the use of cautery, were described in the Edwin Smith Surgical Papyrus (*c.* 3000 BC). Both Hippocrates (400 BC) and Galen (200 AD) recommended its use for accessible tumours. Although tissue destruction by means of cautery is still employed, hyperthermia as we would recognize it today commenced with the observations by Busch in 1866 of tumour regression in two patients developing high fever whilst suffering from erysipelas (Busch 1866). Subsequently William Coley reported cures in patients deliberately inoculated with live streptococci and then with his mixed bacterial toxin preparations (Coley 1893).

Over the next 60 years there were sporadic attempts to heat tumours locally with hot water, high-frequency currents (diathermy) and microwaves. Pioneer work by Warren (1935) demonstrated the feasibility of whole body heating by means of an externally heated cabinet without administration of endogenous pyrogens. Isolated limb perfusion with drugs in a heated perfusate was reported 20 years ago with apparently excellent results in melanoma and sarcoma (Cavaliere et al. 1967, 1980; Stehlin et al. 1979).

Since these earlier clinical experiments considerable renewed interest in hyperthermia has been generated. The extensive experimental studies have been complemented by clinical experience in many thousands of patients using a variety of heating techniques. Hyperthermia has been used alone or, more usually, in combination with radiotherapy. There are also a few studies of its combination with a variety of chemotherapeutic agents.

Hyperthermia is now employed clinically in the treatment of cancer following the induction of tumour temperatures of 40–45 °C. Various methods of heat induction and temperature measurement are used for tumours at different sites. The realization that failure of local tumour control remains a significant contributing factor to both the morbidity and mortality of many cancer patients after conventional radiotherapy has promoted clinical studies of the combined regimes of radiation and hyperthermia.

This chapter reviews the current status of hyperthermia concentrating on clinical techniques and results. Brief consideration will be given to relevant aspects of biology and also to possible future developments. Further details of all aspects of hyperthermia may be obtained from more extensive recent books on the subject (Hahn 1982; Storm 1983; Hornback 1984; Watmough and Ross 1986) and meeting reports (Wizenberg and Robinson 1975; Streffer 1978; Arcangeli and Mauro 1980; Overgaard 1984b).

Biology

The experimental aspects of hyperthermia biology are reviewed in detail in the preceding chapter. The several reasons for its interest as a potential treatment modality are listed in Table 18.1. It should be stressed that these data have been obtained using animal cells and in vivo systems, usually from the mouse. Few data exist about the thermal sensitivity of human cancer cells under experimental conditions. Clinical proof of the enhancement of radiation response is good in the palliative situation of superficial nodules but evidence from curative treatments awaits suitable trials. This likewise applies to proof of an interaction with chemotherapeutic agents in man.

Table 18.1. The potential value of hyperthermia in the treatment of cancer in man: reasons from experimental models

1. Tumour cells are no less sensitive to heat than normal cells
2. Cells at low pH are more sensitive to heat than normal cells. Thus nutritional deprivation in the region of radiobiological hypoxia may be expected to potentiate the effect of heat
3. Cells are resistant to heat and X-rays at different stages of the cell cycle
4. Heat will potentiate the effect of X-rays and some cytotoxic drugs
5. Altered blood flow in tumours may result in temperatures higher in tumour than in normal tissue when localized heating methods are employed. The microvasculature in tumours may also be more sensitive to heat than that in normal tissues

Physiological studies on blood flow and pH changes have been reported but the results are not always consistent with each other (Reinhold and Endrich 1986). Their integration and correlation with the results from experimental models is therefore difficult. These studies are discussed in a later section of this chapter. It is important to note that such physiological factors may have a profound effect on the efficacy of heating. Thus unpredictability of pre-existing and changing blood flow patterns renders uniform heating by external sources difficult. Alterations in metabolic status, including pH, will also change the effectiveness of such therapy during a course of treatment.

Heating Methods

Introduction

A major limitation in the design of hyperthermia treatments suitable for use in man relates to inadequacies in the instrumentation. These result from the difficulty of heating tissues uniformly at a depth of more than 2–4 cm by external means. Only whole body hyperthermia, isolated perfusion and possibly interstitial techniques can achieve such an aim at this time. The limitation is imposed by the physical characteristics of electromagnetic beams, which generally preclude adequate focussing or uniform heating in depth. Ultrasound is marginally more successful but introduces its own problems.

Assessment of the temperatures achieved is also of major importance, particularly in local treatments. This is needed not only to avoid excessive heating which may result in normal tissue damage but also to determine the thermal dose achieved in tumours. All current techniques are invasive, requiring insertion of a temperature detector into the tissue of interest. The requirement to monitor as many points in the tissue of interest as possible increases the burden to the patient and the labour-intensiveness of the therapy. It can be seen, therefore, that in spite of the promising biological reasons for using hyperthermia, current practice is severely limited by the methodology.

A summary of the methods used for inducing hyperthermia in man is presented in Table 18.2. For comprehensive reviews of the techniques from the physics point of view the reader is referred to other sources (e.g. Nussbaum 1982; Hand and James 1986). Brief descriptions of these various methods are given in the succeeding sections. They include whole body heating; regional heating by isolated perfusion or intracavitary infusion; and localized heating by external microwave (MW), radiofrequency (RF) and ultrasound (US) beams or by interstitial applicators.

Whole Body Heating

Whole body heating methods were pioneered by Warren (1935) who used a cabinet in which the patient was enclosed and treated by radiant heat and diathermy. Subsequent developments of this method have included the Siemans–Pomp cabinet (Pomp 1978; Engelhardt et al. 1982; Van der Zee et al. 1983) and a radiant heat device (Robins et al. 1984).

Other techniques, which also require heat transfer through the skin and reduction of skin cooling by impeding sweat evaporation, include the use of hot wax baths (Pettigrew et al. 1974) and the envelopment in a hot blanket (Larkin 1979) or space suit (Bull et al. 1979).

All these techniques require considerable effort, usually needing anaesthesia and with stress to the patient. Local skin burns, especially at pressure points, were not infrequent. Other major systemic complications have included disseminated intravascular necrosis (Pettigrew et al. 1974) and treatment-related death (Van der Zee et al. 1983). Tumour responses following treatment at the maximum tolerated temperature of around 42 °C for 1–2 hours on up to six

Table 18.2. Methods used to induce clinical hyperthermia

Technique	Heating at depth (>4 cm)	Uniformity of tumour temp.	Principal advantages	Principal disadvantages
Whole body Wax bath Heated cabinet Hot blanket Suit Perfusion	Good	Good	Systemic treatment which is not site- or tissue-dependent ± chemotherapy	Limit of 42 °C. May require anaesthesia. No temperature difference between normal tissue and tumour
Isolated perfusion Limb/pelvis	Good	Good	Localized to region ± chemotherapy	Requires open surgery and vascular isolation. No temperature differential
Intracavitary infusion Bladder Peritoneum	Poor	Poor	Localized surface treatment ± chemotherapy	Normal tissue damage. Poor heating below surface
Ultrasound (0.3–5 MHz)	Good	Adequate	Cheap. Can focus single and overlapping beams	Poor air transmission. Reflection at air and bone interfaces. Some interaction with thermocouples
Radiofrequency (10–30 MHz)	Poor	Fair	May heat large volumes. Possible heating at depth. Skin cooling possible	Localization poor. Fat/skin excessively heated with capacitive method. Thermometry may interact
Microwave (30–2450 MHz)	Poor	Fair	High frequencies for superficial lesions. Phased arrays may heat at depth	No true focussing. Large applicators. Thermometry may interact
Interstitial				
Radiofrequency	Difficult	Good	Thermometry can be included as well as brachytherapy	Invasive. Small volume
Coaxial microwave	Difficult	Good		
Ferromagnetic seed	Yes	Unknown	Permanent. Implantable at open surgery at depth	Not yet available

occasions, have been reported with or without concomitant chemotherapy. The transient nature of the responses and the morbidity of the treatment, though, have generally discouraged further investigations. However recent work has been described using a cabinet which does not require patient anaesthesia (Robins et al. 1984). No data are currently available to document its efficacy, and until further information is available whole body hyperthermia cannot be recommended.

Perfusion

Induction of hyperthermia by perfusion with heated blood may be used to induce whole body heating (Parks et al. 1979). More usually it has followed isolation of the arterial and venous supply of a limb in the management of sarcomas and melanomas (Cavaliere et al. 1980; Stehlin et al. 1979). Treatment is combined with chemotherapy. The advantages of this technique are the localization of both controlled temperatures and the drugs to the affected extremity. However, the operation may be associated with considerable morbidity in inexperienced hands, perhaps in part because of the lack of a temperature differential between tumour and normal tissues.

Infusion

Localized intracavitary heating has been attempted by infusion of heated liquids. Treatments of the bladder failed to control tumour and was associated with severe mucosal damage (Ludgate et al. 1976). Likewise, exploration of the technique (together with added chemotherapy) as a method of treatment for secondary deposits from carcinoma of the ovary also failed (Spratt et al. 1980). The technique cannot, therefore, be recommended at this time.

Ultrasound

Ultrasound (US) beams generated at 0.3–5 MHz may be used to heat tissues. They are usually generated by a piezoelectric transducer crystal and the maximum surface area treated by a single beam will depend on the size of this crystal. The beams have a small angle of divergence which permits focussing. Defined volumes at depth may therefore be heated either by using a single transducer in a scanning mode or by employing groups of transducers activated as either annular or phased arrays.

Other advantages of US heating are that there is only a small degree of interaction with the conventional fine metallic thermocouples used for temperature measurement. Also the equipment required is relatively inexpensive compared with that required for radiofrequency and microwave heating. In spite of these several important advantages of US the techniques have not been widely used because of the major disadvantage of US reflection at tissue interfaces, particularly those of bone and air. Differences in attenuation and reflection of US passing through a complex body cross-section will cause major changes in energy distribution. Severe dose-limiting bone pain is thought to be one clinical consequence. The sites at which it can be used are therefore rather limited.

Radiofrequency Beams

Two principal methods are used for coupling the energy of radiofrequency beams (RF) to heat tissues: inductive and capacitive. The most frequently used frequencies are 13.56 and 27.12 MHz. These are the permitted ISM (industrial, scientific and medical) frequencies not normally used for radiocommunication. In addition, a commercial capacitive device (Thermotron RF-8) has recently been developed operating at 8 MHz (Abe et al. 1986).

Capacitive heating occurs when electric fields between pairs of electrodes produce currents. These may be single pairs of electrodes (e.g. Abe et al. 1986) or multiple pairs (e.g. Le Veen et al. 1976). Contour matching with cooled saline bolus bags between the electrodes and skin reduces the risk of superficial heating. Even so, problems may arise with excessive heating of subcutaneous fat due to its high electrical resistance and poor thermal conductivity and vascular perfusion.

Inductive heating occurs when an alternating current in a coil produces a magnetic field. This results in a flow of current which is greatest in adjacent tissues with a high water content such as muscles and viscera. Rapid attenuation means that only relatively superficial tumours can be heated by means of "pancake" coils. Attempts at deeper heating by using a larger coil which encloses a region of the patient (Storm et al. 1979) have not been very successful.

Microwaves

Microwave (MW) radiation in the frequency range 30–2450 MHz is the most commonly used method of inducing hyperthermia. The standard ISM frequencies, not requiring shielded enclosures, are 434, 915 and 2450 MHz.

Many different designs of applicators are used, usually in direct contact with the body surface. Cooling of the superficial tissues is often built into the applicators because maximum power deposition will usually occur close to the surface. The heat distribution depends on the MW frequency, dielectric loading and the shape of the applicator. Heating of superficial masses (up to 4–5 cm depth) may be achieved with single applicators or pairs. A commercially designed BSD 1000 annular phased array of 16 applicators (at 50–100 MHz) concentrically placed around a region of the patient may heat centrally placed tumours as a result of constructive interference (Sapozink et al. 1984). The patient is coupled to the applicators by a series of water bolus bags which also act to cool the superficial tissues. Phantom tests have confirmed the theoretical heating pattern of this equipment and early clinical data also confirm that deep heating can be achieved, particularly in the pelvis (Sapozink et al. 1984, 1985, 1986; Howard et al. 1986b). However, localization of heating to a tumour volume remains difficult although attempts to do so by phase adjustment have been reported (Sathiaseelan et al. 1986; Howard et al. 1986b).

Interstitial and Intracavitary Applicators

As a result of the difficulties in achieving uniform heating by externally applied methods, more direct heating techniques have been investigated. These include the introduction of coaxial MW interstitial applicators into tissues (Cosset et al. 1985) and intraluminal applications into the oesophagus (Sha et al. 1984).

An alternative technique for heating tissue uses resistive RF. This relies on local current flow (LCF) between pairs of implanted electrodes at frequencies in the range of 0.5–1 MHz (Lilly et al. 1983). This technique has been combined with the use of interstitial brachyradiotherapy. The same metallic guides used to introduce the RF electrodes may be used to introduce iridium (^{192}Ir). Thermistors for temperature measurement may also be built into the same electrodes. This technique is clearly likely to prove very useful for tumours accessible to conventional brachytherapy.

An interesting, but as yet clinically unproved technique includes the use of permanent implants of small ferromagnetic seeds (Lilly et al. 1985). These are constructed with a special metallic compound (nickel–copper alloy) such that they can be treated by an externally placed RF field. The Curie point of the seeds is designed so that they heat to a defined temperature (45–47 °C) and no higher. They are thus self-regulating at a defined maximum temperature. The procedure can then be repeated with the same implant if required.

There remain problems with this putative method, principally that ideal ferromagnetic compounds have not yet been defined. Secondly there is the problem that all heat is generated within the implanted metal and has to be conducted away from it to the remaining tumour. In spite of these difficulties the method deserves further research effort.

Thermometry

The need to have adequate control over both tumour and normal tissue temperatures is obvious. Otherwise inadequate tumour heating and excessive normal tissue temperature will lead either to failure of control or excessive tissue damage. Unfortunately, at the present time all available techniques require invasive procedures for insertion of temperature detectors at the appropriate sites.

One problem common to the use of MW and RF heating is that the metallic thermocouples and thermistor leads may heat. This can also happen to a lesser extent during US treatments. Methods to overcome this problem include the use of fine, single or multijunction thermocouples implanted with their long axis as perpendicular to the incident field as possible. The temperature is recorded after temporarily switching off, for a few seconds, the electromagnetic power (Sathiaseelan et al. 1985). Alternative non-perturbing probes include fluorescent sensors (Wickersheim and Alves 1979) or gallium–arsenide birefringent crystals and crystal detectors (Christensen 1977) at the end of optical probes (Cetas 1976), and small semiconductors (Bowman 1976).

Further problems are associated with invasive thermometry. The relatively large overall diameter of the detectors (even the smallest are around 0.5 mm in diameter) will affect local tissue blood flow. Inhomogeneities in tumour heating can only be approximated because of the relatively few detectors which can be positioned. It is also extremely difficult to place detectors in tumours at depth. This may require open surgery, although placement under computer tomographic (CT) guidance can be of use (Howard et al. 1986a).

Several groups have been attempting to develop non-invasive thermometry techniques. These include infrared radiometry and ultrasonic radiometric methods employing US, infrared and MW (reviewed by Christensen in Nussbaum, 1982). None of these techniques has yet achieved anywhere near clinical usefulness.

General Technology Conclusions

The ideal requirements for heating and temperature measurement are summarized in Table 18.3. Current methods clearly lag behind these ideals and are a major restriction to the successful clinical application of treatment.

Table 18.3. Physical requirements for hyperthermia techniques

1. Should heat at whatever depth is required
2. Should be unaffected by tissue inhomogeneities
3. Should treat a defined volume with reasonable temperature homogeneity
4. Should provide adequate thermometry to define a profile of minimum tumour and maximum normal tissue temperatures
5. Equipment should be reliable and easy to use

Clinical Results

The last few years have seen a growth in the use of hyperthermia in the clinic. It is estimated that between 1977 and 1984 11 000 patients were treated with hyperthermia. Most of these will have received simultaneous treatment with other modalities, mainly irradiation or cytotoxic drugs.

Superficial Tumours

Hyperthermia Alone

Some studies have investigated the effect of heat alone on superficial nodules. They have demonstrated an overall response rate of around 50% with a complete response rate of the order of 10%–20% (Table 18.4). The duration of response before tumour regrowth, however, is short at around 6 weeks. These results are remarkably consistent, considering the variables between the different series not only in tumour type and size but also in the technique of heating and temperatures achieved. It would thus appear that as heat alone results in a low complete response rate of short duration it does not have a role in the curative treatment of human tumours. The majority of clinical work now being pursued has therefore concentrated on hyperthermia in combination with other treatment modalities such as radiation or cytotoxic drugs.

Table 18.4. Hyperthermia alone in the treatment of human superficial malignancy

Reference	Objective response rate (%)	Complete response rate (%)
Marmor et al (1979)	54	12
Storm et al. (1979)	50	15
Luk et al. (1981)	36	18
Overgaard (1984a)	51	13

Hyperthermia with Radiation

Many studies have investigated the combination of radiotherapy and superficial hyperthermia (reviewed by Overgaard 1984a). Encouraging anecdotal responses following the combination of these two modalities have been reported for a long time, but results have been difficult to assess. In most reports radiation doses vary considerably, and some studies have employed non-standard fractionation. Intratumour temperatures, both planned and achieved, may vary from treatment to treatment. The positioning of temperature-measuring probes and the number of probes used seem to be critical. In early studies central tumour temperatures were often monitored and one would expect these to be higher than at the periphery. More recent studies have concentrated on measuring the minimum tumour temperature, for reasons discussed later. Many different methods of inducing hyperthermia have been used. Some produce more homogenous heating than others. In addition there are the biological variables of tumour site, size and histology.

Studies incorporating control tumours receiving radiotherapy alone are now available which confirm the synergistic interaction between heat and radiotherapy. Various tumour types have been treated, the majority of patients having superficial nodules of metastatic or primary disease often at sites previously irradiated to high doses. As a result, the radiotherapy dose was often low. A number of studies have demonstrated preferential heating of the tumour compared with normal tissue (Kim et al. 1978). However this does not seem to be a universal finding and may in part be due to inhomogenous power deposition.

These clinical studies are tabulated in Table 18.5. The results are remarkably consistent considering the variables involved. The overwhelming conclusion is that hyperthermia increases the percentage of both the complete and partial responses. Overall response rates increased from around 50% for radiation alone to around 90% for the combined modality, the complete response rate being increased from around 30% to 60%.

Only one study has compared radical courses of radiotherapy and hyperthermia with radical radiotherapy alone (Scott et al. 1984). Radiation doses of 60–66 Gy given in 30–35 fractions over 6 to 6½ weeks were used. The addition of hyperthermia increased the number and speed of responses, and reduced the recurrence rate in 31 matched lesions. Six months after treatment the combined

Table 18.5. Radiotherapy and hyperthermia in the treatment of human superficial malignancy: controlled clinical studies

Reference	Radiation dose (Gy)	RT + HT[a]		RT alone		No. of lesions
		CR (%)[b]	OR (%)[b]	CR (%)	OR (%)	
Kim et al. (1978)	—	88	—	12	—	26
Marmor and Hahn (1980)	Variable	27	87	27	40	30
U et al. (1980	18–42	86	86	14	71	—
Perez et al. (1981)	20–40	69	73	—	—	26
Luk et al. (1981)	15–30	41	78	—	—	37
Kim et al. (1982)	—	—	75	—	46	>100
Arcangeli et al. (1983)	60	76	—	46	—	64
Arcangeli et al. (1983)	40	71	—	43	—	41
Arcangeli et al. (1983)	30	88	—	31	—	33
Scott et al. (1983)	21–24	100	100	—	80	24
Lindholm et al. (1984)	30	—	84	—	48	53
Van der Zee et al. (1984)	13–50	—	90	—	55	113
Bicher et al. (1984)	20–40	63	95	—	—	208
Scott et al. (1984)	60–66	87	—	39	—	31
Overgaard (1984a)	Variable	54	90	32	55	1221
Dunlop et al. (1985)	Variable	74	—	55	—	—
Perez et al. (1986)	20–60	12	—	37	—	164
Howard et al. (1987)	8–54	45	90	33	48	41

[a] RT, radiotherapy; HT, hyperthermia.
[b] CR, complete response; OR, objective response.

modality yielded a complete response rate of 87% compared with 39% for radiotherapy alone, with no increase in local toxicity.

These clinical results seem to confirm the benefit of combining radiotherapy and hyperthermia. However, the end-point used for assessment may be misleading. An increase in overall response rate (complete and partial) demonstrates only that the treatment may be useful for palliation; a more significant end-point is the maintenance of the complete response. It is thus encouraging that the addition of heat to radiation increases the complete response rate in most studies. In the majority of these reports only a short follow-up was possible due to the advanced state of the disease. As a consequence late toxicity and long-term tumour control are not yet adequately investigated.

As more attention has been paid to adequate and accurate thermometry a number of facts have emerged from the clinical studies. Several groups have now confirmed a direct relationship between intratumour temperature and response. The complete response rate has been shown to be higher with increasing average (Van der Zee et al. 1985) and minimum (Oleson et al. 1984; Van der Zee et al. 1986) recorded intratumour temperatures. It now seems fairly certain that the response of a tumour to heat is related to the minimum tumour temperature (Dewhirst et al. 1984). Thus in one series, when some intratumour temperatures reached 42.5 °C the complete response rate was twice that obtained when that temperature was not achieved. However, if the minimum tumour temperature was 42.5 °C the complete response rate was four and a half times greater (Sim et al. 1984). It would appear that toxicity is also related to temperature, with severe normal tissue damage correlating with high maximum recorded temperatures (Dewhirst et al. 1984).

Thus, ideally during a hyperthermia treatment there should be a homogenous temperature rise across the tumour volume. This may account for the important effect tumour size has on response to hyperthermia, either alone or with radiotherapy. Several studies have now shown that therapeutic temperatures are more easily reached and maintained in small tumours (Dunlop et al. 1985; Howard et al. 1987), and almost all published series have demonstrated that small lesions are more likely to achieve a complete response. The results are summarized in Table 18.6. A tumour volume effect has also been shown for radiation alone, and this must be taken into account when analysing the response of lesions treated with either or both modalities.

Table 18.6. The relationship between lesion size and response to combined hyperthermia and radiation

Reference	Lesion size	Response rate (%)
Dewhirst and Sim (1984)	<1.8 cm^3	85 CR[a]
	1.8–8.29 cm^3	55 CR
	8.3–49 cm^3	48 CR
	≥ 49 cm^3	46 CR
Sim et al. (1984)	<25 cm^3	41 CR
	25–125 cm^3	23 CR
	>400 cm^3	0 CR
Perez and Sapareto (1984)	≤ 2 cm	49 CR
	>4 cm	42 CR
Van der Zee et al. (1985)	99.3 cm^3 (mean vol.).	Non-responders
	148.7 cm^3 (mean vol.)	Partial responders
	7.3 cm^3 (mean vol.)	Complete responders
Valdagni et al. (1986)	<6 cm diameter	75 CR
	>6 cm diameter	36 CR
Perez et al. (1986)	<3 cm	80 CR
	>3 cm	65 CR
Howard et al. (1987)	<12 cm^2	57 CR
	>12 cm^2	17 CR

[a] CR, complete response.

Other factors may also affect the response of tumours to combined radiotherapy and hyperthermia. It is hardly surprising that increasing radiotherapy doses have been reported to improve the response rate in many series (Van der Zee et al. 1986). It is reported that various histological tumour types including sarcomas, melanomas, adenocarcinomas and lymphomas may respond better than other tumours (Arcangeli 1984). Recurrent disease and head and neck tumours have also been reported to respond well.

Normal Tissue Response

Of paramount importance is whether the increase in response rate following the addition of hyperthermia to radiotherapy is accompanied by a concomitant increase in normal tissue damage. In man, as in laboratory animal experiments, the normal tissue studied has invariably been skin. Overgaard and Overgaard (1984) have demonstrated in the treatment of melanoma nodules that the enhancement of radiation by heat is maximal if the two are given close together: that is the hyperthermia is given within 30 minutes of the radiotherapy. This thermal enhancement also applies to the skin, however, and there is no resulting therapeutic gain. If the hyperthermia is delayed by 4 hours the thermal enhancement of both tissues is decreased, that in the skin relatively more than in tumour, leading to a therapeutic gain factor of 1.3. This assumes the tumour and

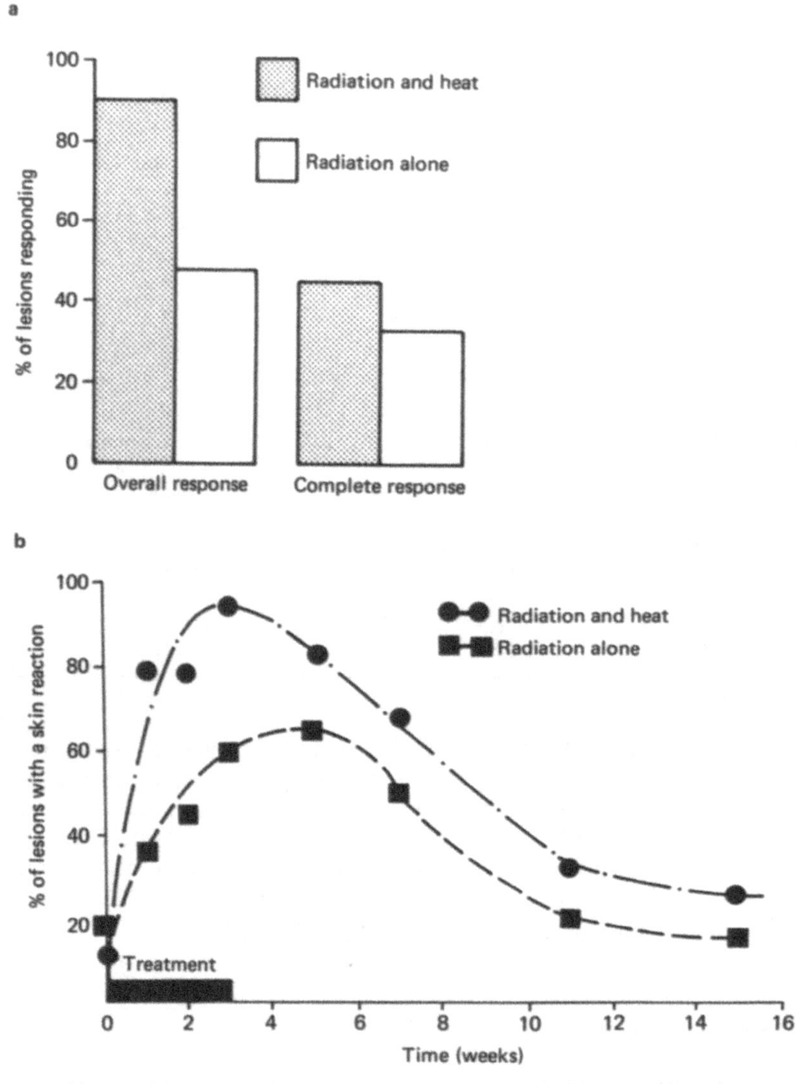

Fig. 18.1. **a** Percentage of lesions treated with hyperthermia and radiation or radiation alone developing a skin reaction during and after treatment. **b** Response of superficial malignant lesions treated with hyperthermia and radiation or radiation alone.

normal tissue are heated to the same degree. A greater therapeutic gain will be achieved if the tumour is heated preferentially because of either an effect of the tumour vasculature or active cooling of normal tissue.

The results of our own study of superficial lesions treated with hyperthermia and radiation tend to confirm many of the factors discussed. The addition of hyperthermia to radiation increased both the complete and the partial response rate. There was also an indication that heat increased the speed of response. This was, however, at the expense of an increase in the severity of skin reactions recorded. A greater proportion of lesions treated with hyperthermia and radiation developed skin reactions than did those treated with radiation alone (Fig. 18.1). The reactions tended to be more severe, to develop more quickly and were related to the maximum skin temperature recorded. This increase in normal tissue toxicity may be related to the relatively high radiation dose per fraction (4 Gy) and the close timing of the two modalities. The hyperthermia was given immediately (within 30 minutes) after the radiotherapy. Both these factors have been shown to increase the skin reaction.

Using multiple different radiotherapy and hyperthermia protocols Arcangeli and his colleagues have observed that thermal enhancement increases with increasing radiotherapy fraction size and heat dose, but at the expense of increased skin reaction (Arcangeli et al. 1983). The greatest therapeutic gain of 2.18 was achieved if hyperthermia was given 4 hours after radiotherapy, compared with a gain factor of 1.14 for simultaneous treatments. Those authors also conclude that thermotolerance does not play a major part in the response of tumours in the clinic.

No studies have so far adequately assessed late normal tissue changes. It has been suggested that the normal tissue damage caused by 50 Gy in 25 fractions over 33 days plus hyperthermia to the breast is less than a radiotherapy dose of 60 Gy in 30 fractions over 40 days given alone (Hofman et al. 1984). This study also suggests that a course of hyperthermia is approximately equivalent to 10 Gy given in five daily fractions, a value which is also reached by Bicher et al. (1984). Such data need confirmation.

To date, therefore, it would appear that both the complete and partial response rate of tumours to radiotherapy can be increased by the addition of hyperthermia. The greatest therapeutic gain is achieved either by preferentially heating the tumour simultaneously with radiotherapy or by delaying the hyperthermia for 4 hours.

Deep-Seated Tumours

Recently developed electromagnetic techniques have now enabled some deep-seated tumours to be heated. The capacitive, inductive and radiative methods of application have all been used.

Le Veen et al. (1976) have used up to three pairs of electrodes applied to the body to induce capacitive heating at depth. A similar technique with specially designed opposed electrodes has also been used in the Thermatron (Abe et al. 1986). This capacitive method of heating may result in preferential heating of subcutaneous fat. Phase I clinical studies appear to confirm this as a limiting factor in both of these devices, except in those patients without much subcutaneous fat. The induction loop method has been developed for regional

Table 18.7. Regional hyperthermia using electromagnetic techniques: results of clinical studies

Reference	Technique	Modalities used[a]	Site treated	No. of patients	Results
Sapozink et al. (1984)	APA (BSD 1000) 60–90MHz	HT + RT	Pelvis	28	Mean heat dose 6.3 min/eq/43
			Abdomen	18	Treatments limited by local and systemic toxicity
Sapozink et al. (1986)	APA (BSD 1000) 60–90 MHz	HT + RT	Abdomen	28	Mean heat dose 3.7 min/eq/43 limited by systemic toxicity
Hornback et al. (1986)	Diathermy 434 MHz	HT + RT	Pelvis	18	Central tumour temps. of 39.5–41.5 °C achieved
Marchal et al. (1985)	HPRL27 system 27 MHz	HT alone or HT + RT	Pelvis and thorax	20	Intratumour temps >42 °C reached in 50% of treatments. Possible to heat chest tumours
Abe et al. (1986)	8 MHz capacitive (Thermatron)	HT + RT (50 Gy)		13 with tumours >6 cm deep	Fat heating a problem. Inadequate thermometry in deep tumours
Howard et al. (1986b)	APA (BSD 1000) 55–65 MHz	HT + RT	Pelvis	20	Pelvic temps. >42 °C reached in 78% of treatments. Heat dose low with 41% >5 min/eq/43

[a] HT, hyperthermia; RT, radiotherapy.

deep heating by Storm as the Magnetrode (Storm et al. 1979, 1982a). The BSD 1000 annular phased microwave array (APA) is now being used by several groups including our own. Details of these techniques have already been discussed in an earlier section. Although each of the methods of regional heating has its own advantages and disadvantages, technical and clinical studies indicate that the APA deposits more power and will raise deep-seated tumours to higher temperatures than a magnetic induction coil. However, the APA may still yield rather unpredictable heating patterns (Paulsen et al. 1985; Sapozink et al. 1985; Oleson et al. 1986).

Clinical studies to date using these methods have all been Phase I trials of feasibility and are summarized in Table 18.7. There are various inherent problems with regional hyperthermia. Whatever the method used there is usually a degree of systemic hyperthermia induced and as a consequence this may result in some cardiovascular stress. Some sites are easier than others. When the pelvis is being heated the APA can maintain a reasonable differential between regional and core temperature. If the region to be heated is situated higher up in the abdomen then a minimal, if any, temperature differential exists (Sapozink et al. 1984; Emami et al. 1984). Thermometry is a problem in that probe insertion in deep-seated tumours may be difficult or impossible. Intracavitary measurements such as in the bladder or vagina are no substitute for true intratumour measurements. CT-guided insertion of thermometry probes (Howard et al. 1986a) and automated mapping devices (Gibbs 1983) allow more extensive temperature measurements.

Clinical results so far are anecdotal. Most patients have received concurrent radiotherapy and/or chemotherapy. Our own experience in Cambridge using the APA is that it has been possible to achieve temperatures in excess of 42 °C in 78% of treatments. However, recorded heat doses have been low. The limiting factors in most cases have been local pain and systemic heating (Howard et al. 1986b). Most toxicity has been acute and restricted to the time of treatment, with no increase in late toxicity being noted. A number of more serious side-effects have been reported from other centres. There have been three cases of reversible unilateral leg weakness with paraesthesiae (Emami et al. 1984), femoral artery rupture, perianal abscess, vaginal vault necrosis and haemorrhage (Sapozink et al. 1984) and gastric ulceration (Hiraoka et al. 1984).

The assessment of response in these patients is virtually impossible in that most have had prior treatment, life-span is short and usually there is another treatment concurrent with the hyperthermia. Most workers, however, have reported palliation of symptoms in the majority of patients treated. Studies so far in regional hyperthermia indicate that this is a technically difficult modality but that therapeutic temperatures can now be reached at depth in the majority of cases. Our initial experience does suggest that further work in patients with less advanced disease is indicated.

Thermal Dose

The concept of a thermal dose is an attractive one. Such a unit allows comparison between different heat treatments, different centres and different parts of the heated volume. Heat dose units have been developed and are now widely used. As temperatures vary with time during a heat treatment the time–temperature relationship must be incorporated into the calculation of a heat dose. Most models use a factor of 2 for a 1 deg C rise in temperature above 42.5 °C, and a factor of between 4 and 8 below this temperature. Most dose units now refer to a minute equivalent at 43 °C (Dunlop et al. 1984; Sapareto 1984).

The biological basis for such a heat unit has been questioned, particularly with regard to multiple treatments, step-up and step-down heating, and combination with other treatment modalities. Opinion regarding the use of such dose units remains divided. However, such a unit would seem to be of value for comparative purposes as long as the limitations of its biological basis are understood.

Interstitial Hyperthermia and Radiotherapy

Increasing interest has developed recently in the use of interstitial methods of inducing hyperthermia. Two methods of LCF and radiative MW applicators are proving to be promising in the clinical situation. The applicators are introduced and a hyperthermia treatment is usually given before and/or after the radioactive implant. Regular geometry of the implant is essential to avoid the development of hot spots. The experience of several centres investigating these techniques is that they result in a homogenous localized temperature rise that is well tolerated by the patient. Temperatures of 44 °C, with little variation across the tumour volume, are possible in the majority of treatments (Cosset et al. 1985; Dunlop et al. 1986). Reported response rates are high, with a 71% complete response rate and a 100% overall response rate recorded in one small series (Manning et al. 1982). No excess toxicity has been noted. However, the benefit of adding hyperthermia to what is a very effective treatment when used alone is still unproven and the results of randomized studies are awaited with interest.

These techniques would seem to be the best currently available for inducing controlled hyperthermia. However, the limitations are obvious in that only sites accessible to and safe for this invasive procedure are suitable. Attempts have been made to heat deeper tumours by implanting one electrode and placing a second externally. This method has been used to treat intrathoracic tumours (Lilly et al. 1983). Clinical experience using implanted ferromagnetic seeds is again limited, but both these methods result in more homogenous heating of deeply situated tumours.

Hyperthermia and Drugs

In contrast to the plethora of studies combining radiation and hyperthermia, there are relatively few clinical studies available on the combination of drugs and hyperthermia. There are inherent problems in studies of this sort. Response rates to the drugs are not well documented and controls for patients treated by drugs alone are essential. A chemotherapy regime makes the availability of hyperthermia-alone controls difficult. In addition to this standard chemotherapy regimes are usually prolonged, with injections of combinations of drugs often 3-weekly, so that courses of hyperthermia are also extended and may be poorly tolerated.

Probably the most encouraging results from combining chemotherapeutic agents and hyperthermia are still those of Stehlin et al. (1979) and Cavaliere et al. (1980). The combination of hyperthermic limb perfusion and melphalan has resulted in impressive 5-year survival figures for patients with locally extensive melanoma. Similarly encouraging, although uncontrolled, results have been achieved by Cavaliere treating a variety of sarcomas and melanomas of the extremities. More recently regional hyperthermic perfusion with melphalan has been used to treat 39 patients with stage I melanoma prior to excision. The 10-year actuarial survival of 86% was significantly better than that of non-randomized controls (Rege et al. 1983). This technique is not without toxicity, with both amputation of the limb and death following some treatments (Cavaliere 1967). In experienced hands, however, it offers the possibility of a homogenous hyperthermia of a whole limb to therapeutic temperatures. Its value awaits the results of randomized clinical trials currently in progress.

Various other techniques have been used to combine hyperthermia and drugs in other tumour sites. Intra-arterial doxorubicin has been combined with hyperthermic bladder perfusion (Jacobs et al. 1981), and melanoma liver deposits have been treated with DTIC and hyperthermia (Storm et al. 1982b). Although these and other reports are encouraging, the number of patients treated is small and controls usually not available.

Some controlled studies do seem to confirm the benefit of combining heat and drugs. Arcangeli and his colleagues have shown a significant increase in response of head and neck nodes treated with hyperthermia and adriamycin or bleomycin over the response to drug alone (Arcangeli et al. 1980). In addition, tumours which appear to be resistant to a chemotherapeutic agent may become sensitive when heat is added. In one series 15% of 34 patients resistant to a variety of drugs demonstrated a response when hyperthermia was combined with the same drug. Over 50% of these previously progressing lesions became stable (Storm et al. 1984). However the result in this small series was not statistically significant.

Whole Body Hyperthermia

The various methods of inducing whole body hyperthermia have already been discussed. All methods are time-consuming and potentially hazardous. Temperatures are limited to less than 42 °C. Above this temperature the morbidity or indeed mortality of the procedure becomes unacceptable. At 42 °C the cell kill from hyperthermia alone will not be great and there is no temperature differential between the tumour and normal tissues. Considering these factors it is perhaps surprising that objective responses have been reported after whole body hyperthermia used alone (Pettigrew et al. 1974).

Most studies have combined this technique with cytotoxic chemotherapeutic agents (Engelhardt et al. 1982; Gerard et al. 1984). Many different drugs have been used to treat a variety of disseminated tumours with variable results. Objective response rates have on occasions been higher than might be expected for chemotherapy alone but data are difficult to interpret as the numbers of patients are small and controls not normally available. A randomized controlled trial has, however, been undertaken by Engelhardt et al. (1982). Patients with small cell carcinoma of the bronchus have been randomized to receive chemotherapy (vincristine, adriamycin and cyclophosphamide) alone or in combination with whole body hyperthermia at 40.5 °C for 1 hour. Fifteen of 30 assessable patients have received the combined modality and as yet no significant survival benefit has been demonstrated. Twenty-nine per cent of the patients who received whole body hyperthermia survived 1 year compared with none who were treated with chemotherapy alone. Further follow-up of these patients has not maintained this survival advantage (R. Engelhardt, personal communication).

Physiological Response to Hyperthermia

Blood Flow

An important reason why hyperthermia may be of use in the treatment of tumours is that differences between tumour and normal tissue blood flow may lead to preferential heating of the former. Tumour neovasculature is morphologically different and the overall vascular density may be reduced compared with that of normal tissues (Gerweck 1984). This does not necessarily result in the tumour blood flow being less than that of normal tissues, as has been shown in animal and human studies (Song 1984; Beaney et al. 1984). Tumour vasculature, however, does not respond to a heat stress as effectively as normal tissue vessels. Several animal studies demonstrate that tumour blood flow does not increase to the same degree as that of normal tissue after hyperthermia. In one series normal tissue flow increased by a factor of more than 6 compared with a factor of 2 for tumour flow (Dudar et al. 1984). In addition, vascular damage and stasis appear to occur at a lower temperature in tumours (41 °C for 60 minutes) than in normal tissues (47 °C for 60 minutes) (Dudar et al. 1984). These physiological changes have also been demonstrated in spontaneous canine tumours, where a minimal increase in flow was seen within mast cell tumours following hyperthermia (Milligan and Danjepour 1985). This differential response between normal and tumour vasculature may result in preferential heating of the tumour. Some human clinical studies have confirmed this, but it is not a universal finding (Kim et al. 1978).

pH and Oxygen Tension

Changes in pH and oxygen tension (pO_2) consequent on the vascular effects may also be important. The extracellular pH within experimental tumours has been shown to be lower than in normal tissues, with levels of around 7.0 or even lower being found (Urano et al. 1980). Measurements made in human tumours of various histological cell types mostly confirm intratumour pH values to be lower than those in normal subcutaneous tissues (Wike-Hooley et al. 1984). Changes in tumour vasculature following heat, and the subsequent reduction in perfusion, may further reduce the pH resulting in increased thermosensitivity, but data are conflicting and there is some evidence that pH increases following hyperthermia (Thistlethwaite et al. 1985). Clinical work in human tumours has shown that pO_2 and blood flow initially increase on heating to 41 °C, but above this both decrease. With multiple heat treatments this initial increase is abolished and the pO_2 and flow stabilize below the pre-heat values (Bicher and Mitagvaria 1984).

Conclusions

Laboratory studies clearly demonstrate the potential of hyperthermia both alone and in combination in the treatment of cancer. Clinical studies have lagged far behind their laboratory counterparts. The reason for this is that presently available heating techniques are in general not good enough. Heat delivery by external means is unreliable in all but the smallest superficial tumours, and interstitial methods, although promising, have yet to be proven. Thermometric techniques are also inadequate. Although many exciting non-invasive methods are being investigated at present, the only techniques routinely used are invasive. It may be that clinical hyperthermia cannot fulfil its potential until there are technical improvements in both heat delivery and thermometry.

Clinical studies have now confirmed that the addition of heat to radiation increases both partial and complete response rates, and probably the speed and duration of response. By suitable timing of the two procedures or cooling normal tissues a therapeutic gain can be achieved. The place of localized hyperthermia in the clinic is still unknown, however, as the sites which can be treated are limited by currently available techniques.

Results presently available from studies using regional hyperthermia are anecdotal. Randomized studies are under way. If the increased response rate seen in superficial tumours can be demonstrated in deep tumours, where local control is a problem, there may well be a useful role for this modality. Specific sites such as advanced neck nodes in head and neck cancer, and chest wall recurrences from cancer of the breast, seem particularly promising. Hyperthermia has not yet fulfilled its initial promise, but it has already been shown to be of some value in the clinic and warrants further clinical study.

References

Abe M, Hiraoka M, Takahashi M et al. (1986) Multi-institutional studies on hyperthermia using an 8 MHz radiofrequency capacitive heating device (Thermatron RF-8) in combination with radiation for cancer therapy. Cancer 58:1589–1595

Arcangeli G (1984) Hyperthermia and radiation, biological and clinical studies. In: Overgaard J (ed) Proceedings of the 4th international symposium on hyperthermic oncology, vol 2. Taylor and Francis, London, pp 283–293.

Arcangeli G, Mauro F (eds) (1979) Proceedings of the 1st meeting of the European Group of Hyperthermia in Radiation Oncology, Cambridge, 9–10 September 1979. Masson, New York

Arcangeli G, Cividalli A, Lovislo G et al. (1980) Effectiveness of local hyperthermia in association with radiotherapy or chemotherapy: comparison of multimodality treatments on multiple neck node metastases. In: Arcangeli G, Maura F (eds) Proceedings of 1st meeting of the European Group of Hyperthermia in Radiation Oncology, Cambridge, 9–10 September 1979. Masson, New York, pp 257–265

Arcangeli G, Cividalli A, Lovisolo G, Nervi C (1983) Clinical results after different protocol of combined local heat and radiation. Strahlentherapie 159:82–89

Beaney RP, Lammertsa A, Jones T, McKenzie CG, Halnan KE (1984) Positron emission tomography for in vivo measurement of regional blood flow, oxygen utilization and blood volume in patients with neck cancer. Lancet I:131

Bicher H, Mitagvaria N (1984) Changes in tumour tissue oxygenation during microwave hyperthermia: clinical relevance. In: Overgaard J (ed) Proceedings of the 4th international symposium on hyperthermic oncology, vol 1. Taylor and Francis, London, p 169

Bicher H, Wolstein R, Fingerhut A, Lewinsky B, Frey H (1984) Clinical multifield controlled comparison of hyperthermia and low dose radiation to full dose radiation of direct wall recurrences. In: Overgaard J (ed) Proceedings of the 4th international symposium on hyperthermic oncology, vol 1. Taylor and Francis, London, p 39

Bowman RR (1976) A probe for measuring temperature in radio-frequency heated material. IEEE Trans MTT 24:43–45

Bull JM, Lees D, Schuette W et al. (1979) Whole body hyperthermia: a Phase I trial of a potential adjuvant to chemotherapy. Ann Intern Med 90:317–323

Busch W (1886) Verhandlungen ärtlicher Gesellschaften. Berl Klin Wochenschr 3:245–246

Cavaliere R, Crocatto E, Giovanella B et al. (1967) Selective heat sensitivity of cancer cells. Cancer 20:1351–1381

Cavaliere R, Moncea G, Di Filippo F, Caputo A (1980) Heat transfer problems during local perfusion in cancer treatment. Ann NY Acad Sci 335:311–326

Cetas TC (1976) A birefringent crystal optical thermometer for measurement of electromagnetically induced heating. In: Biological effects of electromagnetic waves, vol 2. Selected papers of the USNC/URSI meeting, Boulder, Colorado, pp 338–348

Christensen DA (1977) A new nonperturbing probe using semiconductor band edge shift. J Bioeng 1:541–545

Coley WB (1893) The treatment of malignant tumors by repeated inoculations of erysipelas: with a report of ten original cases. Am J Med Sci 105:487–511

Cosset JM, Dutreix J, Haie C, Gerbaulet A, Janoray P, Dewar JA (1985) Interstitial thermoradiotherapy: a technical and clinical study of 29 implantations performed at the Institut Gustave-Roussy. Int J Hypertherm 1:3–13

Dewhirst MW, Sim DA (1984) The utility of thermal dose as a predictor of tumour and normal tissue responses in combined radiation and hyperthermia. Cancer Res 44 [Suppl]:4772s–4780s

Dudar TE, Jain RK (1984) Differential response of normal and tumour microcirculation to hyperthermia. Cancer Res 44:605–612

Dunlop P, Dickinson R, Hand J, Field S (1984) The use of thermal dose in the clinical application of localised hyperthermia. In: Overgaard J (ed) Proceedings of the 4th international symposium on hyperthermic oncology, vol 1. Taylor and Francis, London, p 187

Dunlop PRC, Dickinson RJ, Hand JW (1985) A study of thermal dose responses of human tumours using local hyperthermia. Strahlentherapie 161:531

Dunlop PRC, Dickinson RJ, Hand JW, Munro AJ, Vallis KA (1986) Early experience with combined interstitial hyperthermia and brachytherapy. Br J Radiol 59:525–527

Emami B, Perez C, Nussbaum G, Leybovich L (1984) Regional hyperthermia in the treatment of recurrent deep-seated tumours: preliminary report. In: Overgaard J (ed) Proceedings of the 4th international symposium on hyperthermic oncology, vol 1. Taylor and Francis, London, p 605

Engelhardt R, Neumann H, van der Tarn M, Löhr GW (1982) Preliminary results in the treatment of oat cell carcinoma of the lung by combined application of chemotherapy and whole body hyperthermia. Prog Clin Biol Res 107:761–765

Gerard H, van Echo DA, Whiteacre M et al. (1984) Doxorubicin, cyclophosphamide, and whole body hyperthermia for treatment of advanced soft tissue sarcoma. Cancer 53:2585–2591

Gerweck L (1984) Environmental and vascular effect. In: Overgaard J (ed) Proceedings of the 4th

international symposium on hyperthermic oncology, vol 2. Taylor and Francis, London, p 325
Gibbs FA (1983) Thermal mapping in experimental cancer treatment with hyperthermia: description and use of semiautomatic system. Int J Radiat Oncol Biol Phys 9:1057–1063
Hahn G (ed) (1982) Hyperthermia and cancer. Plenum Press, New York
Hand JW, James JR (eds) (1986) Physical techniques in clinical hyperthermia. Wiley, Chichester, pp 1–558
Hiraoka M, Jo S, Takahashi M, Abe M (1984) Treatment of locally advanced hepatocellular carcinoma with RF hyperthermia. In: Overgaard J (ed) Proceedings of the 4th international symposium on hyperthermic oncology, vol 1. Taylor and Francis, London, p 803
Hofman P, Lagendijk J, Schipper J (1984) The combination of radiotherapy with hyperthermia in protocolised clinical studies. In: Overgaard J (ed) Proceedings of the 4th international symposium on hyperthermic oncology, vol 1. Taylor and Francis, London, p 379
Hornback NB (ed) (1984) Hyperthermia and cancer: human clinical trial experience, vols 1 and 2. CRC Press, Boca Raton, Florida
Hornback NB, Shupe RE, Shidma H, Marshall CU, Lauer T (1986) Advanced stage IIIB cancer of the cervix treatment by hyperthermia and radiation. Gynecol Oncol 23:160–167
Howard GCW, Dixon AK, Bleehen NM (1986a) Computed tomography guided cannula insertion in pelvic hyperthermia. Br J Radiol 59:860
Howard GCW, Sathiaseelan V, King GA, Dixon AK, Anderson A, Bleehen NM (1986b) Regional hyperthermia for extensive pelvic tumours using an annular phased array applicator: a feasibility study. Br J Radiol 59:1195–1201
Howard GCW, Sathiaseelan V, Freedman L et al. (1987) Hyperthermia and radiation in the treatment of superficial malignancy: an analysis of treatment parameters, response and toxicity. Int J Hypertherm (in press)
Jacobs SC, Maddison FE, Lawson RK (1981) Doxorubicin hypogastric artery infusion combined with hyperthermia therapy for transitional cell carcinoma of the bladder. Cancer Treat Rep 6:891–893
Kim JH, Hahn EW, Tokita N (1978) Combination hyperthermia and radiation therapy for cutaneous malignant melanoma. Cancer 41:2143–2148
Kim JH, Hahn EW, Ahmed SA (1982) Combination hyperthermia and radiation therapy for malignant melanoma. Cancer 50:478–482
Larkin JM (1979) A clinical investigation of total-body hyperthermia as cancer therapy. Cancer Res 39:2252–2254
Le Veen HH, Wapnick S, Piccone V, Falk G, Ahmed N (1976) Tumour eradication by radiofrequency therapy. J Am Med Assoc 235:2198–2224
Lilly MB, Brezovich IA, Atkinson W et al. (1983) Hyperthermia with implanted electrodes: in vitro and in vivo correlations. Int J Radiat Oncol Biol Phys 9:373–382
Lilly MB, Brezovich IA, Atkinson WJ (1985) Hyperthermia induction with thermally self-regulating ferromagnetic implants. Radiology 154:243–244
Lindholm CE, Kjellen E, Landberg T, Nilsson P, Persson B (1984) Microwave induced hyperthermia and radiotherapy: clinical results. In: Overgaard J (ed) Proceedings of the 4th international symposium on hyperthermic oncology, vol 1. Taylor and Francis, London, p 341–344
Ludgate CM, McClean N, Carswell GF, Newsam JE, Pettigrew RT, Selby Tulloch W (1976) Hyperthermic perfusion of the distended urinary bladder in the management of recurrent transitional cell carcinoma. Br J Urol 47:841–848
Luk KH, Purser PR, Castro JR, Meyer TS, Phillips TL (1981) Clinical experiences with local microwave hyperthermia. Int J Radiat Oncol Biol Phys 7:615–619
Manning MR, Cetar TC, Miller RC, Oleson JR, Connor WG, Gerner EW (1982) Results of a phase I trial employing hyperthermia alone or in combination with external beam or interstitial radiotherapy. Cancer 49: 205–216
Marchal C, Bey P, Jacomino JM, Hoffstetter S, Garland ML, Robert J (1985) Preliminary technical experimental and animal results of the use of the HPLR 27 system for the treatment of deep- seated tumours by hyperthermia. Int J Hypertherm 1:105–116
Marmor JB, Hahn GM (1980) Combined radiation and hyperthermia in superficial human tumours. Cancer 46:1986–1991
Marmor JB, Pounds D, Postic TB, Hahn GM (1979) Treatment of superficial human neoplasms by local hyperthermia induced by ultrasound. Cancer 43:188–197
Milligan AJ, Danjepour M (1985) Canine normal and tumour tissue estimated blood flow during fractionated hyperthermia. Int J Radiat Oncol Biol Phys 11:1679–1684
Nussbaum GH (ed) (1982) Physical aspects of hyperthermia. American Institute of Physics, New York
Oleson JR, Sim DA, Manning MR (1984) Analysis of prognostic variables in hyperthermia treatment of 161 patients. Int J Radiat Oncol Biol Phys 10:2231–2239

Oleson JR, Sim DA, Conrad J, Fletcher A, Gross EJ (1986) Results of a phase I regional hyperthermia device evaluation: microwave annular assay versus radiofrequency induction coil. Int J Hypertherm 2:327–336
Overgaard J (1984a) Rationale and problems in the design of clinical studies. In: Overgaard J (ed) Proceedings of the 4th international symposium on hyperthermic oncology, vol 2. Taylor and Francis, London, p 325
Overgaard J (ed) (1984b) Hyperthermic oncology, vols 1 and 2. Taylor and Francis, London
Overgaard J, Overgaard M (1984) A clinical trial evaluating the effect of simultaneous or sequential radiation and hyperthermia in the treatment of malignant melanoma. In: Overgaard J (ed) Proceedings of the 4th international symposium on hyperthermic oncology, vol 1. Taylor and Francis, London, pp 383–386
Parks LC, Minaberry D, Smith DP et al. (1979) Treatment of far-advanced bronchogenic carcinoma by extra-corporeally induced systemic hyperthermia. J Thorac Cardiovasc Surg 78:883–892
Paulsen KD, Stronbehn JW, Lynch DR (1985) Comparative theoretical performance for two types of regional hyperthermia systems. Int J Radiat Oncol Biol Phys 11:1659–1671
Perez CA, Sapareto S (1984) Thermal dose expression in clinical hyperthermia and correlation with tumour response/control. Cancer Res 44 [Suppl]:4818s–4825s
Perez CA, Kopecky W, Ven Kata Roa D, Baglan R, Mann J (1981) Local microwave hyperthermia and irradiation in cancer therapy: preliminary observations and directions for future clinical trials. Int J Radiat Oncol Biol Phys 7:765–772
Perez CA, Kuske RR, Emami B, Fineberg B (1986) Irradiation alone or combined with hyperthermia in the treatment of recurrent carcinoma of the breast in the chest wall: a randomised comparison. Int J Hypertherm 2:179–187
Pettigrew RT, Galt JM, Ludgate CM, Smith AN (1974) Clinical effects of whole-body hyperthermia in advanced malignancy. Br Med J iv:679–682
Pomp H (1978) Clinical application of hyperthermia in gynecological malignant tumours. In: Streffer C, van Beuningen D, Dietzel F et al. (eds) Cancer therapy by hyperthermia and radiation. Urban and Schwarzenberg, Baltimore, pp 326–327
Rege VB, Leone LA, Soderberg CH et al. (1983) Hyperthermic adjuvant perfusion chemotherapy for stage I malignant melanoma of the extremity, with literature review. Cancer 52:2033–2039
Reinhold HS, Endrich B (1986) Tumour microcirculation as a target for hyperthermia. Int J Hypertherm 2:111–137
Robins HI, Neville AJ, Shecterle LM et al. (1984) Clinical trials of a radiant heat whole body hyperthermia system. In: Overgaard J (ed) Proceedings of the 4th international symposium on hyperthermic oncology, vol 1. Taylor and Francis, London, pp 269–272
Sapareto S (1984) Thermal dose determination in cancer therapy. Int J Radiat Oncol Biol Phys 10:781–801
Sapozink MD, Gibbs FA, Gates K, Stewart R (1984) Regional hyperthermia in the treatment of clinically advanced deep-seated malignancy: results of a pilot study employing an annular assay applicator. Int J Radiat Oncol Biol Phys 10:775–786
Sapozink MD, Gibbs FA, Thompson JW, Ettingham JR, Stewart JR (1985) A comparison of deep regional hyperthermia from an annular phased study and a corrective coil in the same patients. Int J Radiat Oncol Biol Phys 11:179–190
Sapozink MD, Gibbs FA, Egger MJ, Stewart JR (1986) Abdominal regional hyperthermia with an annular phased array. J Clin Oncol 4:775–783
Sathiaseelan V, Howard GCW, Har Kedar I, Bleehen NM (1985) A clinical microwave hyperthermia system with multipoint real-time thermal dosimetry. Br J Radiol 58: 1187–1195
Sathiaseelan V, Iskander MF, Howard GCW, Bleehen NM (1986) Theoretical analysis and clinical demonstration of the effect of power pattern control using the annular phased array hyperthermia system. IEEE Trans MTT 34:514–519
Scott RS, Johnson RJ, Kowal H, Kushnamsetty RM, Story K, Clay L (1983) Hyperthermia in combination with radiotherapy: a review of five years experience in the treatment of superficial tumours. Int J Radiat Oncol Biol Phys 9:1327–1333
Scott RS, Johnson RJ, Story K, Gay L (1984) Local hyperthermia in combination with definitive radiotherapy: increased tumour clearance and reduced recurrence rate in extended follow-up. Int J Radiat Oncol Biol Phys 10:2119–2123
Sha Y-H, Li D-J, Qiu S-L, Hou F-X, Li Z-C, Zhao Y-Z (1984) The combined treatment of esophageal cancer – a clinical pathological study of 42 cases. In: Overgaard J (ed) Proceedings of the 4th international symposium on hyperthermic oncology, vol 1. Taylor and Francis, London, pp 371–374
Sim D, Oleson J, Grochowski K (1984) An update of the University of Arizona human clinical

hyperthermic experience including estimates of therapeutic advantage. In: Overgaard J (ed) Proceedings of the 4th international symposium on hyperthermic oncology, vol 1. Taylor and Francis, London, p 367

Song CW (1984) Effect of local hyperthermia on blood flow and microenvironment: a review. Cancer Res 44 [Suppl]:4721s–4730s

Spratt JS, Adcock RA, Muskovin M, Sherrill W, McKeown J (1980) Clinical delivery system for intraperitoneal hyperthermic chemotherapy. Cancer Res 40:256–260

Stehlin JS, Giovanella BC, de Ipolyi Anderson RT (1979) Eleven years' experience with hyperthermic perfusion for melanoma of the extremities. World J Surg 3:305–307

Storm FK (ed) (1983) Hyperthermia in cancer therapy. GK Hall, Boston

Storm FK, Harrison W, Elliott R, Morton D (1979) Normal tissue and solid tumour effects of hyperthermia in animal models and clinical trials. Cancer Res 39:2245–2251

Storm FK, Elliott RS, Harrison WH, Morton DL (1982a) Clinical RF hyperthermia by magnetic loop induction: a new approach to human cancer therapy. IEEE Trans MTT 30:1149–1157

Storm FK, Kaiser LR, Goodnight JE et al. (1982b) Thermochemotherapy for melanoma metastases in liver. Cancer 49:1243–1248

Storm FK, Silberman AW, Ramming KC et al. (1984) Clinical thermochemotherapy: a controlled trial in advanced cancer patients. Cancer 53:863–868

Streffer C, van Beuringen D, Dietzel F et al. (eds) (1978) Cancer therapy by hyperthermia and radiation. Proceedings of the 2nd international symposium, Essen, West Germany, 2–4 June 1977. Urban and Schwarzenberg, Baltimore

Thistlethwaite AJ, Leeper DB, Moylan DJ, Nerlinger RE (1985) pH distribution in human tumours. Int J Radiat Oncol Biol Phys 11:1647–1652

U R, Noell KT, Woodward KT, Warden BT, Fishburn RI, Miller RS (1980) Microwave-induced local hyperthermia in combination with radiotherapy of human malignant tumours. Cancer 45:638–646

Urano M, Gerweck L, Epstein R, Cunningham M, Suit H (1980) Response of a spontaneous murine tumour to hyperthermia: factors which modify the thermal response in vivo. Radiat Res 83:312–322

Valdagni R, Kapp DS, Valdagni C (1986) N_3 (TNM–UICC) metastatic neck nodes managed by combined radiation therapy and hyperthermia: clinical results and analysis of treatment parameters. Int J Hypertherm 2:189–200

Van der Zee J, Van Rhoon GC, Wike-Hooley JL et al. (1983) Whole-body hyperthermia in cancer therapy: a report of a phase I–II study. Eur J Cancer Clin Oncol 19:1189–1200

Van der Zee J, Van Rhoon GC, Wike-Hooley JL, Van den Berg AP, Reinhold HS (1984) Thermal enhancement of radiotherapy in breast carcinoma. In: Overgaard J (ed) Proceedings of the 4th international symposium on hyperthermic oncology, vol 1. Taylor and Francis, London, pp 345–348

Van der Zee J, Van Rhoon GC, Wike-Hooley JL (1985) Clinically derived dose effect relationship for hyperthermia given in combination with low dose radiotherapy. Br J Radiol 58:243–250

Van der Zee J, van Putten WLJ, Van den Berg AP et al. (1986) Retrospective analysis of the response of tumours in patients treated with a combination of radiotherapy and hyperthermia. Int J Hypertherm 2:337–349

Warren SL (1935) Preliminary study of the effect of artificial fever upon hopeless tumor cases. Am J Roentgenol 33:75–87

Watmough DJ, Ross WM (eds) (1986) Hyperthermia. Blackie, London pp 1–245

Wickersheim KA, Alves RV (1979) Recent advances in optical temperature measurement. Ind Res Dev 21:82–89

Wike-Hooley JL, Haveman J, Reinhold HS (1984) The relevance of tumour pH to the treatment of malignant disease. Radiother Oncol 2:343–366

Wizenberg MJ, Robinson JE (eds) (1975) Proceedings of the international symposium on cancer therapy by hyperthermia and radiation, Washington DC, 28–30 April 1975. American College of Radiology, Washington DC

19 Monoclonal Antibodies in the Diagnosis and Treatment of Cancer

L. W. Brady, D. V. Woo, A. M. Markoe, H. Koprowski and C. T. Miyamoto

Introduction

The identification of the monoclonal antibody technique by Kohler and Milstein (1975), as well as advances in immunological methods (Koprowski et al. 1978), have led to steadily improved techniques for producing antibodies that target antigens on human tumours in vivo. The developments have included radioimmunoassay and immunohistochemistry methods that make it possible to identify discrete tumour antigens (Bast et al. 1981, 1984; Del Villano et al. 1983).

It is possible to use the monoclonal antibody method as a tool to identify a host of tumour-associated antigens in human tumours (Koprowski et al. 1978; Ross et al. 1986). These antigens serve as a target for the radiolabelled antibody (Moldofsky et al. 1983; Chatel et al. 1984; Halpern et al. 1985) and for some tumours several suitable antigen/antibody systems have been discovered. Further, the monoclonal antibody method offers a major technical advance in that large amounts of homogeneous, well-characterized immune-specific antibodies can be readily produced in a sterile and pyrogen-free form to meet an ever-increasing demand. Of principal importance is their use for diagnosis of human malignant tumours, as well as for treatment of tumours either alone or armed with anti-tumour substances such as radioactive isotopes (Order et al. 1980), chemotherapeutic agents or toxins (Bast et al. 1984).

Tumour-specific heterosera (polyclonal antibodies) obtained from animals immunized against a given tumour were labelled with radioactive iodine for the purpose of diagnostic imaging of tumours and demonstrated the feasibility of such an approach. The first studies were performed in animals and showed that improved tumour targeting could be achieved with these heterosera. Initial successes in imaging human tumours were reported with heterosera against carcinoembryonic antigen (CEA), human chorionic gonadotrophin, alpha-fetoprotein, kidney carcinoma and ferritin (Belitsky et al. 1978; Goldenberg et al. 1978; Dykes et al. 1980; Mach et al. 1980; Order et al. 1980a,b). The early reports cited diagnostic accuracies for metastatic deposits in the order of 85%.

Radiolabelled heterosera were also used in therapeutic studies in humans. Order et al. (1980a,b) used iodine-131-labelled anti-ferritin IgG and iodine-131-labelled anti-CEA IgG as a part of an integrated treatment programme for primary hepatic malignancies. About two thirds of patients treated with the radiolabelled antibody preparation showed tumour regression. The results were, however, influenced by the fact that the patients received regimens including chemotherapy and external radiation therapy to the liver as well as the radioactive labelled antibody.

Selection of an antibody for studies in humans depends on a variety of factors such a molecular size (IgG versus antibody fragments), reactivity after labelling, metabolic clearance and mass dose. Antibodies that have been successful as tumour imaging agents have usually reacted with normal tissue with less than 1% of their reactivity with tumour tissue (Fawwaz et al. 1985).

Monoclonal antibodies are promising indicators of disease and have potential therapeutic applications, given their higher homogeneity and specificity than polyclonal antibodies. These antibodies can combine with cell-killing drugs or radioactivity and target specific tumour cells which express the appropriate antigenicity. Clinical studies have demonstrated the ability of certain radiolabelled monoclonal antibodies to localize in specific tumours in vivo, yielding nuclear images of the tumour (Ballou et al. 1979; Mach et al. 1983; Chatel et al. 1984; Halpern et al. 1985; Hnatowich et al. 1985). However, the extent of localization is not always as specific as initially predicted from animal model studies in nude mice. This decrease in antibody specificity can be attributed to many factors but it has been difficult to assess these problems because of the uniqueness of the antibodies, each of which has individual physicochemical characteristics. The basis for any diagnostic imaging or radiation therapy approach using protein carriers such as monoclonal antibodies, therefore, depends upon their ability to demonstrate a selectivity of tumour over normal tissues.

Although the methods for producing tumour-specific monoclonal antibodies, whether mouse or human immunoglobulins, are advancing rapidly, the techniques for modifying the antibody to carry radioactivity or drugs are still a major topic for research. Inappropriate modifications of certain monoclonal antibodies have resulted in an attenuation of their specificity for tumour tissue. It is therefore necessary to evaluate the possible approaches to chemical modification of monoclonal antibodies and define areas that require greater effort, as well as to consider the potential radiobiological questions resulting from the presumed effects of low dose rates (Hall et al. 1986). The areas on which research has been focussed as regards the use of monoclonal antibodies in the diagnosis and therapy of cancer are discussed in the subsequent sections of this chapter.

Selection of Radionuclides

The selection of an appropriate radionuclide requires consideration of factors such as photon versus particle emission for imaging or therapy; the required energy to be employed for imaging or therapy; and the bio-compatibility of the isotope from a chemical point of view. Some of the appropriate radionuclides are listed in Table 19.1.

Table 19.1. Selected radionuclides for diagnosis and therapy with monoclonal antibodies

(a) For diagnosis

Radionuclide	Usable photons for imaging		
	Half-life	Energy (keV)	% intensity
^{67}Cu	2.57 d	93	16.7
		184	47.7
^{67}Ga	3.261 d	93	32.3
		185	19.7
		300	16.0
^{131}I	8.040 d	284	6.1
		364	81.2
		637	7.3
^{123}I	13.13 h	159	83.4
^{111}In	2.83 d	171	90.2
		245	94.0
^{47}Sc	3.4 d	159	68.0
^{117m}Sn	13.6 d	159	86.4
^{99m}Tc	6.02 h	140	89.0

(b) For therapy

Radionuclide	Primary particulate emissions			
	Half-life	Type, energy (MeV)	% intensity	g-rad/μCi-h
^{211}At	7.214 h	Alpha, 5.87	41.7	5.21
^{212}Bi	60.45 min	Alpha, 6.05	25.2	3.25
^{67}Cu	61.88 d	4 betas, 0.141 E_{av}	100.0	0.301
^{131}I	8.04 d	5 betas, 0.181 E_{av}	99.9	0.387
^{32}P	14.29 d	1 beta, 0.695 E_{av}	100	1.48
^{188}Rc	16.98 h	7 betas, 0.764 E_{av}	100	1.63
^{47}Sc	3.44 d	2 betas, 0.162 E_{av}	100	0.345
^{90}Y	64.1 h	1 beta, 0.934 E_{av}	99.98	1.99

From Kocher (1981).

The use of radionuclide imaging is the first key step in the evaluation of any monoclonal antibody that is targeted towards specific tumours (Ballou et al. 1979; Powe et al. 1984; Fawwaz et al. 1985). The selection of the optimal radionuclide for imaging is based upon its particular radiochemistry and current limitations of nuclear medicine instrumentation. The physical decay parameters of a given radionuclide are also contributing factors in the selection procedure. Radionuclides with relatively high photon abundances within the low to medium energy range (100–200 keV) are necessary for efficient detection with conventional sodium iodide crystals employed in present-day cameras. On this criterion iodine-123 and technetium-99m are the most suitable. However, their physical half-lives are too short (<13 hours) to allow for sufficient accumulation of antibody in the tumour – a process which occurs over 2–3 days. Radionuclides which have more favourable half-lives (2–8 days) may not have the ideal photon energies (>250 keV) for their efficient detection. Iodine-131, indium-111 and gallium-67 fall in this category. In general, most diagnostic studies requiring imaging are at present done using indium-111-labelled antibodies.

The selection of appropriate radionuclides for therapy is a question of their relative radiobiological effectiveness. Most radionuclides considered as possible therapeutic isotopes emit radiations having high linear energy transfer (LET). Alpha-emitters such as astatine-211 and bismuth-212 have been reported as possible radionuclides to be combined with monoclonal antibodies for therapeutic applications. High-energy beta-emitters such as scandium-47 and yttrium-90 are also likely candidates for therapy. Radionuclides that decay by electron capture resulting in Auger electrons (iodine-125 and bromine-77) may give intense localized radiation within the intracellular volume of target tissue (Bradley et al. 1975). Other pure electron emitters such as the isotopes of phosphorus (^{32}P) offer potentially great advantages in terms of giving a reasonable radiation dose to target tissues if the appropriate chemical attachment to monoclonal antibodies can be developed. In general, radioimmunotherapeutic approaches in our institution have been carried out using iodine-125-labelled compounds.

Selection of the Functional Modifying Component

Chemical modification of tumour-specific monoclonal antibodies has allowed firm attachment of radionuclide metal atoms resulting in improved in vivo stability and greater tumour uptake (Hnatowich et al. 1982; Meares and Goodwin 1984; Gobuty et al. 1985). Studies using indium-111-labelled 19-9 antibody against colorectal cancer demonstrated tumour-specific uptake in patients with metastatic disease (Hnatowich et al. 1985; Keenan et al. 1985). However, certain chemical modifications of an antibody to improve its ability to carry foreign substances have the potential to cause significant loss of its targeting ability for tumours (Gobuty et al. 1985). The very site-specific feature for which the antibody was selected is easily compromised by the chemical reactions required to couple radionuclides and drugs. It is well recognized that the more groups that are attached to the antibody the greater is its loss of immunospecificity (Lindmo et al. 1985; Fawwaz et al. 1985; Paik et al. 1985).

Selection of Monoclonal Antibodies

While work continues on the development of tumour-specific monoclonal antibodies, it is still unclear as to what constitutes the ideal antibody for carrying tumoricidal radionuclides. Heterogeneity of tumour antigens remains a major obstacle since, ideally, antibody should be delivered to every tumour cell.

Nevertheless it may be possible to develop panels of antibodies with known cross-reactivities to different types of tumours that will allow greater tumour uptake.

Radiochemical Stability

The stability of the monoclonal antibody in vivo is another factor in the selection of appropriate antibodies for radioimmunotherapy. While it is not clear how individual monoclonal antibodies are metabolically degraded, striking differences have been found in the clearance kinetics of different radiolabelled monoclonal antibodies from various organs and tissues, suggesting possible qualitative differences in metabolism (Halpern et al. 1981; Keenan et al. 1985; Khaw et al. 1986).

Radiobiological Considerations

Order and his colleagues have shown definite benefit of iodine-131-labelled polyclonal anti-ferritin therapy in hepatocellular carcinoma at very low dose rates (0.05 Gy/hour or less) (Leichner et al. 1981, 1983). This contrasts with the usually accepted ineffectiveness of low-dose-rate irradiation. It has been postulated by Hall et al. (1986) that dose rate may not be the critical parameter for slowly proliferating cells but that the dose per cell cycle may be more significant.

Current Programme at Hahnemann University

Table 19.2 lists the monoclonal antibodies being investigated in the current programme at Hahnemann University. The majority of the work at present is directed towards the use of the antibodies derived from human colon cancer (17-1A, 19-9 and GA73-3). These are labelled with indium-111 for diagnostic purposes and iodine-125 for therapeutic programmes.

Table 19.2. Monoclonal antibodies currently being investigated at Hahnemann University

1. Human colon cancer derived monoclonal antibodies
   - 17-1A
   - 19-9
   - GA73-3
2. Breast cancer derived monoclonal antibody BR55-2
3. Epidermal growth factor receptor antibody EGFr 425

Age Karnofsky Tumor stage	STRATIFICATION	Hepatic irradiation (15 Gy/10 fractions, 2 weeks)	Leukapheresis Dose escalation MoAB	Weekly X6 Biweekly X6

Reassessment at 6-week intervals:
- Clinical examination
- Chest X-ray
- Blood studies (CBC, SMA-12, CEA)
- CT liver

Fig. 19.1. Pilot study 86–04 for carcinoma of the colon with metastases limited to the liver: monoclonal antibody research protocol.

Table 19.3. Pilot study 86–04 for carcinoma of the colon with metastases limited to the liver: monoclonal antibody leukapheresis summary

Number of patients	14
Average age	63.3 years
Age range	43–78 years
Average number of treatments	6
Average total dose	1185 mg
Average follow-up	4.86 months
Number alive	7
Number deceased	7
Patient population	
Colon metastatic to liver or recurrent	7
Colon with local failure	1
Gastric	1
Lung	2[a]
Vaginal metastatic to liver	1
Pancreatic	1
Acinic cell of parotid gland	1
Response	
Complete response	2[b]
Partial response	1[c]
Progression of disease	4
Deceased	7

[a] One mesothelioma and one adenoma.
[b] Both colon.
[c] Colon.

Age Karnofsky Tumor site	S T R A T I F I C A T I O N	Primary treatment: Surgery ± radiation therapy	Iodine-125 MoAB	IV Day 01, 08, 14 and Repeat monthly X3
		Reassessment monthly Appropriate studies		

Fig. 19.2. Pilot study 86–05 for adjuvant evaluation of monoclonal antibodies labelled with radionuclides: research protocol.

The major protocols currently being used are as follows:

1. The human colon cancer derived antibodies, primarily 17-1A, are being investigated for their potential to identify metastatic disease from primary lesions of the gastrointestinal tract. Indium-111-labelled 17-1A shows a high degree of localization capability in areas of metastatic disease.

2. Epidermal growth factor antibody (EGFr 425) labelled with indium-111 is being used to identify tumours within the brain which expresses EGF receptor. Subsequently this antibody may be labelled with iodine-125 and delivered intra-arterially to treat recurrent primary malignant brain tumours showing localization.

3. The use of 17-1A in the treatment of metastatic lesions involving the liver is being assessed. Fig. 19.1 details this particular protocol and the criteria for patient selection. The results of treatment thus far are presented in Table 19.3.

4. Iodine-125-labelled 17-1A is being used as an adjuvant to surgery and/or definitively administered radiation therapy. This protocol is illustrated in Fig. 19.2. Table 19.4 details the results achieved so far.

Summary

Studies with monoclonal antibodies in the diagnosis and treatment of cancer illustrate the fact that better chemical methods for the attachment of specific radionuclides to the antibodies are necessary to ensure greater stability in vivo. The studies thus far have demonstrated that significant quantities of monoclonal antibodies can be administered over protracted periods of time in multiple fractions without significant complications in their delivery or the appearance of allergic reactions.

Table 19.4. Pilot study 86–05 for adjuvant evaluation of monoclonal antibodies labelled with radionuclides: monoclonal antibody CO17-1A/I-125 summary

Number of patients	24
Average age	57.3 years
Age range	34–76 years
Number of treatments	3.5
Average total dose	18.7 mCi
Average follow-up	2 months
Number alive	15
Number deceased	9
Patient population	
Colon metastatic to liver	9
Lung	2[a]
Maxillary sinus	1
Pancreatic	6
Breast	5
Unknown	1
Response	
Complete response	2
Partial response	1
Stable	5
Progression of disease	7
Deceased	9

[a] One large cell and one small cell carcinoma.
[b] Maxillary sinus and breast.
[c] Large cell carcinoma of lung.

Indium-111-labelled antibody has been found to be an excellent tool for identifying the tracking areas of metastatic disease in patients who have carcinoma of the colon. It can also be used effectively in treatment as an adjuvant to definitively administered surgery and/or radiation therapy.

Cancer imaging with monoclonal antibodies represents a new technique that is broadly applicable to many solid human tumours. By combining the antibody with the proper radionuclide it is possible, for some antibody/tumour systems, to achieve detection of in vivo metastases in about 85% of patients. Current limitations to the technnique include the need to develop radiolabelled preparations with better and more rapid targeting to tumour and less non-specific localization in non-tumour sites such as the liver. Nonetheless cancer imaging with monoclonal antibodies is likely to form an important part of the diagnostic armamentarium of the future. Of equal importance is the fact that by conjugating the antibodies with different radionuclides there is the potential for effective anti-tumour therapy for many tumours against which current therapies are ineffective.

References

Ballou B, Levin G, Hakala TR, Solter D (1979) Tumor location detected with radioactively labeled monoclonal antibody and external scintigraphy. Science 206:844–847

Bast RC, Feeney M, Lazarum H, Nadler LM, Calvin RB, Knapp RC (1981) Reactivity of a monoclonal antibody with human ovarian carcinoma. J Clin Invest 68:133

Bast RC, Klug TL, Schaetzl E et al. (1984) Monitoring human ovarian carcinoma with a combination of CA–125, CA 19–9 and carcinoembryonic antigen. Am J Obstet Gynecol 199:553–559

Belitsky P, Ghose T, Aquino J, Tai J, MacDonald AS (1978) Radionuclide imaging with metastases from renal cell carcinoma by I-131-labeled antitumor antibody. Radiology 126:515–517

Bernstein A, Hurwitz E, Maron R, Arnon R, Sela M, Wilchek M (1978) Higher antitumor efficacy of daunomycin when linked to dextran: in vivo and in vitro studies. J Natl Cancer Inst 60:379–383

Bradley EW, Chan PC, Adelstein SJ (1975) The radiotoxicity of iodine-125 in mammalian cells. 1. Effects on the survival curve of radioiodine incorporated into DNA. Radiat Res 64:555–563

Chatel JF, Saccavini JC, Fumoleau P et al. (1984) Immunoscintigraphy of colon carcinoma. J Nucl Med 25:307–314

DelVillano BC, Brennan S, Brock P et al. (1983) Radioimmunometric assay for a monoclonal antibody defined tumor marker-carbohydrate antigen 19-9. Clin Chem 29:548–552

Dykes PW, Hine KR, Bradwell AR et al. (1980) Localization of tumor deposits by external scanning after injection of radiolabeled anticarcinoembryonic antigen. Br Med J 280:220–222

Fawwaz RA, Wang TST, Estabrook A et al. (1985) Immunoreactivity and biodistribution of Indium-111-labeled monoclonal antibody to a human high molecular weight melanoma-associated antigen. J Nucl Med 26:488–492

Gobuty AH, Kim EE, Weiner RE (1985) Radiolabeled monoclonal antibodies: radiochemical pharmacokinetic and clinical challenges. J Nucl Med 26:546–547

Goldenberg DM, Deland F, Kim E (1978) Use of radiolabeled antibodies to carcinoembryonic antigen for the detection and localization of diverse cancers by external photoscanning. N Eng J Med 298:1384–1388

Hall EJ, Marchese M, Astor M, Morse T (1986) Response of cells of human origin, normal and malignant, to acute and low dose rate irradiation. Int J Radiat Oncol Biol Phys 12:655–659

Halpern SE, Stern PH, Hagan P L et al. (1981) Radiolabeling of monoclonal antitumor antibodies: comparison of I-125 and In-111 CEA with Ga-67 in a nude mouse–human colon tumor model. Clin Nucl Med 6:453

Halpern SE, Dillman RO, Witztum KF et al. (1985) Radioimmunodetection of melanoma utilizing In-111 96.5 monoclonal antibody: a preliminary report. Radiology 155:493–499

Hnatowich DJ, Griffin TW, Koscciuczyk C et al. (1985) Pharmacokinetics of an Indium-111 labeled monoclonal antibody in cancer patients. J Nucl Med 26:849–858

Hnatowich DJ, Layne WW, Childs RL (1982) The preparation and labeling of DTPA-coupled albumin. Int J Appl Radiat Isot 33:327–331

Keenan AM, Harbert JC, Larson SM (1985) Monoclonal antibodies in nuclear medicine. J Nucl Med 26:531–537

Khaw BA, Cooney J, Edgington T, Strause HW (1986) Differences in experimental tumor localization of dual-labeled monoclonal antibody. J Nucl Med 27:1293–1299

Kocher DC (1981) Radioactive decay data tables. Technical information centre, US Department of Energy, DOE/TIC 11026

Kohler G, Milstein G (1975) Continuous cultures of fused cells secreting antibody of predefined specificity. Nature 256:495–497

Koprowski H, Steplewski Z, Herlyn D (1978) Study of antibodies against human melanoma produced by somatic cell hybrids. Proc Natl Acad Sci USA 75:3405–3409

Leichner PK, Klein JL, Garrison JB et al. (1981) Dosimetry of I-131-labeled anti-ferritin in hepatoma: a model for radioimmunoglobulin dosimetry. Int J Radiat Oncol Biol Phys 7:323–333

Leichner PK, Klein JL, Siegelman SS, Ettinger DS, Order SE (1983) Dosimetry of I-131-labeled antiferritin in hepatoma: specific activities in the tumor and liver. Cancer Treat Rep 67:647–658

Lindmo T, Boven E, Cuttitta F et al. (1984) Determination of the immunoreactive fraction of radiolabeled monoclonal antibodies by linear extrapolation to binding at infinite antigen excess. J Immunol Methods 72:77–89

Mach JP, Carrel S, Forni M, Ritschard J, Donath A, Alberto P (1980) Tumor localization of radiolabeled antibodies against carcinoembryonic antigen in patients with carcinoma. N Engl J Med 303:5–10

Mach JP, Chatal JF, Lumbroso JD et al. (1983) Tumor localization in patients by radiolabeled monoclonal antibodies against colon carcinoma. Cancer Res 43:5593–5600

Meares CF, Goodwin DA (1984) Linking radiometals to proteins with bifunctional chelating agents. J Protein Chem 3:215–228

Moldofsky PJ, Powe J, Mulhern CB et al. (1983) Metastatic colon carcinoma detected with radiolabeled F(ab′), monoclonal antibody fragments. Radiology 149:549–555

Order SE, Ettinger JL, Alderson P, Siegelman S, Leichner P (1980a) Phase I–II study of radiolabeled antibody integrated in the treatment of primary hepatic malignancies. Int J Radiat Oncol Biol Phys 6:703–710

Order SE, Klein JL, Ettinger D, Alderson P, Siegelman S, Leichner P (1980b) Use of isotopic immunoglobulin in therapy. Cancer Res 40:3001–3007

Paik CH, Hong JJ, Ebbert MA, Heald SC, Reba C, Eckelman WC (1985) Relative reactivity of DTPA, immunoreactive antibody-DTPE conjugates toward indium-111. J Nucl Med 26:482–487

Powe J, Pak CH, Steplewski Z et al. (1984) Labeling monoclonal antibodies and $F(ab')_2$ fragments with (In-111) Indium using cyclic DTPA anhydride and their in vivo behavior in mice bearing human tumor xenografts. Cancer Drug Deliv 1:125–135

Ross AA, Herlyn D, Iliopoulos D, Koprowski H (1986) Isolation and characterization of a carcinoma-associated antigen. Biochem Biophys Res Comm 135:297–303

Woo DV, Koprowski C, Brady LW, Dadparvar S, Steplewski Z, Koprowski H (1986) Imaging metastatic brain tumors using monoclonal antibodies. J Nucl Med 27:1028

Subject Index

Accelerated fractionation 67, 68
Acute lymphatic leukaemia (ALL) 38, 96
Adenocarcinoma 219
Adenoid cystic carcinoma 146
Adriamycin 205, 225, 226
Alkylating agents 205
Amphotericin B 205
Anaemia 159, 168, 173
Ankylosing spondylitis 48
Anti-cancer drugs, *see* Chemotherapy, *and under specific drugs*
Astatine-211 236

Bismuth-212 236
Bladder cancer 68
Bleomycin 193–5, 205, 225
Blood flow 201, 210
  in hyperthermia 226, 227
Blood pressure 170, 172
Blood vessels, radiation damage 11
Bone marrow
  irradiation effects
    fractionated doses to whole volume 99–101
    single doses to whole volume 99
  regeneration (BMR) patterns 100
  T cell depletion of 125
  transplantation (BMT) 95, 97
Bowel damage 88
Bragg peak 130, 131, 141
Brain tolerance unit (BTU) 38
Brain tumours 126
BrdU 69
Breast fractionation trial 32, 34
Bromine-77 236
Bronchial carcinoma 170, 226
BUdR 26
Buthionine sulphoximine (BSO) 5, 157

Caesium-137 88
Californium-192 84
Cataract formation 115, 126
Cell Adhesive Matrix 7
Cell killing and radiosensitization 178–9
Cell populations within tumours 9–10
Cell radiosensitivity 10
Cellular radiation biology 6–7
Cellular radiation response 1–9
Cellular radiosensitivity 73
Cervical cancer 1, 42, 50, 66, 87, 88, 159, 172
Chemotherapy
  combined with hyperthermia 224–6
  combined with radiotherapy, *see* Combined radiotherapy and chemotherapy
  interaction with hyperthermia 205
Chest X-ray changes 36–7
CHO cells 180, 181
Cisplatin 194
  effects on proliferation 183–4
  effects on repair of radiation damage 180–2
  effects on split dose repair 180
  effects on X-ray repair in vitro 182
  in combined radiotherapy and chemotherapy 177–99
  inhibition of SLD repair by 182
  interaction with X-rays 179
  killing and radiosensitization 178–9
  radiosensitization effects 181
  radiosensitization of RIF1 cells by 178
Cisplatin–DNA adducts 184
CNS irradiation effects, sequelae of 104

CNS syndrome 96–7
CNS toxicity, clinical measurement of 38–9
Colonic adenocarcinoma 138
Colonic carcinoma 238
Combined radiotherapy and chemotherapy 177–99
  clinical studies 193–5
  experimental tumour studies 177–84
  future directions 195–6
  multi-drug chemotherapy regimens 194–5
  normal tissue studies 184–93
  side-effects of 195
Computerized tomography (CT) 36–7, 39, 47, 103
Coronary artery disease 105–7
Creatinine clearance 111
Cumulative radiation effect (CRE) 17, 54, 56
  correction 88, 89
Cyclophosphamide 226
Cytotoxic drugs 36, 165
  as radiosensitizers 193–4
Cytotoxic therapy 38

Deuterons 129
2,3-Diphosphoglycerate 160
DNA adduct 196
DNA damage 5, 19
DNA synthesis 26, 184
Dose–effect curve 33, 67
Dose enhancement factor (DEF) 185
Dose equivalence calculations 89
Dose-rate effect 73–9, 88
  human tumour cell lines 75–7
  mathematical models 75
  normal tissues 77
Dose-reduction factor (DRF) 76
Dose–response curves 33, 36, 39, 48, 57, 67, 69, 70, 103, 189, 190
Dose–time relationships 47, 59–60
Doxorubicin 225
Drug response 18
DTIC 225
Duodenal damage 184–7
Dyspigmentation 48

ED_{50} value 77, 104
Electron-affinic sensitizers 140, 156
Etanidazole 154
EULIMA (European light-ion medical accelerator) project 141
European Organization for Research and Treatment of Cancer (EORTC) 43, 68
Ewing's sarcoma 95
Excisional assays 43
Eye irradiation effects 115

Fibrosis 104
Fitzfractionation 68
Fixed cervical nodes 146
5-Fluorouracil 193, 194
Follicle stimulating hormone (FSH) levels 126
Fractionation radiotherapy 3, 13, 17, 23–4, 34
  accelerated 67, 68, 138
  adjustment of total dose 56–7
  alternative schemes 61–3
  biological factors 60–1
  clinical trials 59–71
    criteria of 63–5
  conventional 61, 65
  dose per fraction 54–5
  factors affecting early reactions 55–6
  factors affecting late reactions 56
  Fitzfractionation 68
  low-dose-rate 90
  non-standard 56–7
  overall time effect 53–4
  predictive radiosensitivity testing 69–70
  principles of 53–8
  split course irradiation 66, 68
  see also Hyperfractionation; Hypofractionation

Gastrointestinal syndrome 96–7
Gastrointestinal tract irradiation effects 111–12
General Health Diary Card 43
Glioblastoma 86
γ-Glutamyl cysteine synthetase 157
Glutathione (GSH) levels 157
Granulocytopenia 99

Haematological response following total body exposure 99
Haemoglobin levels 168–9, 173
Haemoglobin–oxygen affinity 160
Haemoglobin status 159
Haemopoietic syndrome 96–7
Hahnemann University, monoclonal antibodies 237–9
Head and neck cancer 67, 68, 84, 138, 144–6, 167, 168, 171–3, 219
Heart attacks 107
Heart disease, development of 107
Heart irradiation effects 105–8
Heavy ion therapy 140, 141
Helium ions 140
Hepatoma cells 181
Histological assessments 42
Hodgkin's disease 107, 108, 195
^{3}HTdR labelling index 20
Hydralazine 160, 162
Hydroxyurea 193
Hyperbaric oxygen 66, 153, 159, 166–8, 171
  radiotherapy in 172–3
Hyperfractionation 24, 59, 63, 65, 67, 68, 138
Hyperthermia 165
  as cell killing modality 201
  biology 201–8, 210

blood flow in 226, 227
cell response to 202–3
clinical methods and results 209–31
clinical results 216–23
combined with chemotherapy 224–6
combined with radiotherapy 217–19, 224
deep-seated tumours 221–3
effects of 27, 36
effects of sequence and interval 206
heating methods used 211
history of clinical applications 209
hyperthermia alone 216–17
interactions with chemotherapy 205
interactions with radiotherapy 205–6
interstitial methods of inducing 224
normal tissue response 219–21
oxygen tension in 226–7
pH effects in 226–7
physical requirements 216
physiological response to 226–7
potential value in treatment of cancer 210
rationale for use of 201–2
re-treatment of previously irradiated tissue 206
superficial tumours 216–19
thermal dose concept 223–4
thermal dose quantitation 206–7
thermometry techniques 215–16, 218
tissue response to 18, 202–3
whole body 211–16, 225–6
Hypofractionation 24, 63, 65–6
Hypoxia 153
and tumour cell kinetics 171
animal models 169
before radiotherapy 172–3
exploitation for increasing efficiency of radiosensitization 161–2
induction by vasoactive drugs 160–1
manipulation for therapeutic gain 159–62
protective effect of 5
radiobiological 167–72
radioprotectiveness of 167
radiotherapy under 172
role of 11
Hypoxic cell sensitizers 153
multiple mechanism approach 155–9
Hypoxic cells 73
animal models 169
bioreductive activation 156–7
chemical radiosensitizers 166
histological evidence of 170–1
radioresistance of 165, 205
radiosensitization of 165–6
reoxygenation of 168
responsibility for failure 171–2
Hypoxic radiosensitization 179–80

Indium-111 239, 240
Infusion, induction of hyperthermia by 213
Interstitial applicators 215
Interstitial implants 82, 83
Interstitial pneumonitis 96
Interstitial radiotherapy 84–6
Intestinal adenocarcinoma 126
Intracavitary applicators 215
Intracavitary irradiation of gynaecological malignancies 83, 86–9
Iodine-123 235
Iodine-125 84, 86, 236, 239
Iodine-131 234, 237
Iridium-192 85
Isobolograms 179
Isoeffect curves 82–4
Isoeffective doses, calculation of 57–8
IUdR 26

Karnofsky performance status 43
KHT sarcoma 162
KHT tumour 158, 161
Kidney
combined radiotherapy and chemotherapy 188–91
fractionation studies 24, 25
irradiation effects 109–10

Large intestine, combined radiotherapy and chemotherapy 187–8
Late effects 47–52
factors affecting 56
non-stochastic 52
patients at high risk of 70
stochastic 52
total body irradiation 126
LD_{50} values 18–19
$LD_{50/30}$ values 99, 102
$LD_{50/30-60}$ values 97
$LD_{50/100}$ values 99
$LD_{50/160}$ values 102
Lethality studies 18–19
Leukaemia 48, 78, 95, 123, 126
Lewis lung tumour 160
Linear energy transfer (LET) 73, 84, 131, 140
Liver irradiation effects 116
Local control 39–40, 137–8, 153, 193
Low-dose-rate irradiation
clinical implications of 78–9
external beam therapy 84, 89
in clinical radiotherapy 81–93
local field 84
radiobiological considerations 74–5, 81
Lung cancer 195
Lung function effect 169
Lung irradiation effects 36–7
doses to partial volume 103–4
fractionated doses to whole volume 102–3
single doses to whole volume 101
Lymphoma 123, 219
Lymphoproliferative disorders 126

Magnetic resonance imaging 47
Malignant melanoma 146
Marshallese Island 51
Medulloblastoma 38
Megavoltage X-rays 47
Melanoma 219, 220, 225
Methotrexate 205
Metronidazole 156, 166, 167
Microwave radiation, induction of hyperthermia by 214
Misonidazole 5, 153, 156–8, 162, 167, 168, 170–1, 173
Mitomycin C 194
Mitotic death 19–20
Mitotic delay 22–3
Mitotic index 22
Monoclonal antibodies 233–42
  functional modifying component 236
  Hahnemann University programme 237–9
  radiobiological considerations 237
  radiochemical stability 237
  selection of radionuclides 234–6
  selection of tumour-specific 236–7

Neoplastic cells, proliferation kinetics 10–11
Neuroblastoma 95
Neutron therapy 129, 138–40, 143–9
  current status of 144–8
  development of 143–4
  future of 149
  therapeutic index for high- and low-energy 148–9
Nitroimidazole 166
Nitrosoureas 193, 205
Nominal standard dose (NSD) 17, 54, 56, 60
Non-Hodgkin's lymphoma 95, 96, 126, 195
Normal cell tissue radiation response 70
Normal tissue damage, clinical scales of 32
Normal tissue end-points
  application of 33–6
  characteristics of 31–3
  graded scales 33–4
  trial design 33–4
Normal tissue radioprotectors 17
Normal tissue response
  combined chemotherapy and radiotherapy 184
  hyperthermia 219–21
  measurement of 31–45
  reporting of 34–5
  reviews of topics relating to 26
  timing of 34–5
  to radiation 17–29
  total body irradiation 95–121
Normal tissues
  dose-rate effect 77
  dose tolerance of 59
Nuclear magnetic resonance spectroscopy 69
Nuclear reactor accidents 95
Nuclear warfare 95, 99
Nuclear weapons testing 51

Oral cavity implants 83
Oral mucositis 35
Oral tumours 85
Oropharyngeal tumours 90
Orthovoltage X-rays 47
Ovarian cancer 195
Ovary irradiation effects 114
Oxic cells 12, 73, 74
Oxic radiosensitization 179–80
Oxygen effect 4–6, 68, 140
  exploitation of 172–3
Oxygen enhancement ratio (OER) 5, 131, 140
Oxygen–haemoglobin dissociation curve 172
Oxygen tension in hyperthermia 226–7
Oxygen transport effects 172

Pancarditis 105
Paranasal sinus tumours 145–6
Particle therapy 129–35
  appropriate control arm in studies of 138
  characteristics of charged particles 129
  clinical aspects of 137–51
  clinical reality of 139
  depth–dose distribution 130
  evaluation of 138–9
  high LET 140
  intermediate LET 140
  LET distributions 131–2
  low LET 140
  physical aspects of 130–1
  radiobiological effects at high LET 132–3
  radiobiology of 131–3
  rationale for use of 139–40
  use of term 129
Patient response, measurement of 43
Patient self-assessments 43
Pelvic radiation tolerance 88
Pelvic tumours 147
Perfluorochemical agents 172
Perfusion, induction of hyperthermia by 213
Pericarditis 105, 107, 108
pH effects 202, 210
  in hyperthermia 226–7
Phase sensitivities 21
Photon therapy 138, 139
Pimonidazole 154–5, 158
Pion therapy 129, 131, 140–1
cis-Platinum 205
Potentially lethal damage (PLD) 9, 180, 181
  recovery from 11, 13
Predictive assays 6–7, 69–70
Prostate carcinoma 147–8
Proton therapy 129, 139, 140, 142–3
Pulmonary toxicity, clinical measurement of 36–7

Radiation, cellular response to 1–8
Radiation carcinogenesis 48–52
Radiation damage

blood vessels 11
non-stochastic 48
recovery from 1–4, 6
stochastic 48
Radiation lethality syndromes 95
prominent examples of 96–7
Radiation myelopathy 38
Radiation pneumonitis 101
Radiation resistance 73–9
Radiation response
end-points of 13–14
modification of 153–64
of tumours 9–15
Radiation risk estimates 51
Radiation sensitizers 153–64, 173
differential uptake 158–9
dual function 157–8
in clinical radiotherapy 165–75
Radical radiotherapy 59, 61
Radioactive iodine 51
Radiobiology, practical applications of 23–6
Radiofrequency beams, induction of hyperthermia by 214
Radionuclides, selection for diagnosis and therapy with monoclonal antibodies 234–6
Radioprotectors 27
Radiosensitizers 5, 26–7
Radiotherapy
combined with chemotherapy, *see* Combined radiotherapy and chemotherapy
combined with hyperthermia 217–19, 224
interaction with hyperthermia 205–6
Relative biological effect (RBE) 84, 131–2
Renal damage
combined radiotherapy and chemotherapy 188
development of 192
Renal function, combined radiotherapy and chemotherapy 188
Reoxygenation 13
Repair theory 3
RIF1 cells 178–81, 183, 184
Ro03–8799 173
RSU1069 156–61

Salivary gland tumours 145
Sarcoma F tumour 169
Sarcoma 147, 171, 219, 225
Scandium-47 236
Second cancers 195
see also Radiation carcinogenesis
Second tumours 126
Sensitizer enhancement ratio (SER) 5, 178–9
Skin damage 89
Skin fractionation studies 24
Skin response in hyperthermia 220–1
Skin tumours 126, 82
Small cell anaplastic lung cancer 95
Small cell carcinoma 226
Small intestine, combined radiotherapy and chemotherapy 184–7
Source–skin distance (SSD) 89
Split course irradiation 66, 68, 90
Squamous cell cancer 67, 144–5
SR2508 173
Step-down sensitization 204
Sublethal radiation damage (SLD) 180, 181
recovery from 3
repair of 9, 13
Survival curves 1–4, 73, 74, 76, 77, 185–7
for established human tumour cell lines 6
Surviving fraction 67

T cell depletion of bone marrow 125
T cell lymphoma 160
Technetium-99m 235
Telangiectasia 48
Testis irradiation effects 112–13
Testosterone levels 126
Thermal enhancement ratio (TER) 205, 206
Thermometry in clinical hyperthermia 215, 218
Thermotolerance 203–4
in vivo and in vitro 204
molecular basis for 204
operational model 204
reviews 204
time course of 204
Thiols
as radioprotectors 27
suppression of 157
Thoracic irradiation, low-dose-rate 102
Thrombocytopenia 99
Thyroid gland irradiation effects 108–9
Thyroid tumours 51
Thyrotropic hormone (TSH) 108–9
Thyroxine 108
Time dose factor (TDF) 54, 56, 84
Tissue culture
lethal effects of radiation on mammalian cells in 1
practical application of 6–7
Tissue response, cellular basis of 23
Tongue implants 86
Total body irradiation (TBI)
biological effects of 95
clinical aspects 123–7
clinical problems 125
dosimetry 124
fractionated schedule 125
late effects 126
normal tissue effects 95–121
radiation dose 123–5
single-fraction 124, 125
therapeutic application of, 95
toxicity effects 125–6
see also under specific organs
Tourniquet hypoxia 27
Tumour cell kinetics 69, 171
Tumour cell radiosensitivity 6

Tumour control at primary site 137
Tumour control probability 5
Tumour geometry 11
Tumour radiocurability 12
Tumour radiosensitizers 17
Tumour response
  biochemical assays 42
  effect of different modalities 36
  excisional assays 43
  histological assessments 42
  local control 39–40
  measurement of 31–45
  volume growth delay 40
Tumour shrinkage, measurements of 40–2
Tumour vasculature 11
Tumours
  cell populations within 9–10
  radiation response of 9–15

Ultrasound 47
  induction of hyperthermia by 213–14
United Nations Scientific Committee on the Effects of Atomic Radiation (UNSCEAR) 48

Vascular architecture 18
Vascular perfusion 170
Vincristine 226

Whole body hyperthermia 211–16, 225–6
Whole organ tolerance, single versus fractionated doses 98
WR2721 27

Yttrium-90 236

GPSR Compliance

The European Union's (EU) General Product Safety Regulation (GPSR) is a set of rules that requires consumer products to be safe and our obligations to ensure this.

If you have any concerns about our products, you can contact us on ProductSafety@springernature.com

In case Publisher is established outside the EU, the EU authorized representative is:

Springer Nature Customer Service Center GmbH
Europaplatz 3
69115 Heidelberg, Germany

Batch number: 10755110

Printed by Printforce, the Netherlands